Anatomy at a Glance

Third edition

Omar Faiz

Bsc (Hons), FRCS (Eng), MS
Senior Lecturer & Consultant Surgeon
St Mary's Campus
Imperial College, London

Simon Blackburn

BSc (Hons), MBBS, MRCS (Eng)
Specialty Registrar in Paediatric Surgery

David Moffat

VRD, MD, FRCS
Emeritus Professor of Anatomy
Cardiff University

WILEY-BLACKWELL

A John Wiley & Sons, Ltd., Publication

This edition first published 2011 © 2011 by Omar Faiz, Simon Blackburn and David Moffat

Blackwell Publishing was acquired by John Wiley & Sons in February 2007. Blackwell's publishing program has been merged with Wiley's global Scientific, Technical and Medical business to form Wiley-Blackwell.

Registered office: John Wiley & Sons, Ltd, The Atrium, Southern Gate, Chichester, West Sussex, PO19 8SQ, UK

Editorial offices: 9600 Garsington Road, Oxford, OX4 2DQ, UK
The Atrium, Southern Gate, Chichester, West Sussex, PO19 8SQ, UK
111 River Street, Hoboken, NJ 07030-5774, USA

For details of our global editorial offices, for customer services and for information about how to apply for permission to reuse the copyright material in this book please see our website at www.wiley.com/wiley-blackwell

Library of Congress Cataloging-in-Publication Data

Faiz, Omar.
  Anatomy at a glance / Omar Faiz, Simon Blackburn, David Moffat. – 3rd ed.
    p. ; cm. – (At a glance)
  Includes index.
  ISBN 978-1-4443-3609-2
  1. Human anatomy–Outlines, syllabi, etc. I. Blackburn, Simon, 1979- II. Moffat, D. B. (David Burns) III. Title. IV. Series: At a glance series (Oxford, England)
  [DNLM: 1. Anatomy. QS 4]
  QM31.F33 2011
  611–dc22
                                                                    2010029199

A catalogue record for this book is available from the British Library.

Set in 9/11.5pt Times by Aptara® Inc., New Delhi, India
Printed in Singapore by Ho Printing Singapore Pte Ltd

1 2011

Anatomy at a Glance

Companion website

This book is accompanied by a companion website:

www.wiley.com/go/anatomyataglance

The website includes:

- 100 interactive flashcards for self-assessment and revision

Some figures in this book have been reproduced from Diagnostic Imaging, by P. Armstrong, M. Wastie and A. Rockall (9781405170390) © Blackwell Publishing Ltd.

Contents

Preface to the first edition

The study of anatomy has changed enormously in the last few decades. No longer do medical students have to spend long hours in the dissecting room searching fruitlessly for the otic ganglion or tracing the small arteries that form the anastomosis round the elbow joint. They now need to know only the basic essentials of anatomy with particular emphasis on their clinical relevance and this is a change that is long overdue. However, students still have examinations to pass and in this book the authors, a surgeon and an anatomist, have tried to provide a means of rapid revision without any frills. To this end, the book follows the standard format of the *at a Glance* series and is arranged in short, easily digested chapters, written largely in note form, with the appropriate illustrations on the facing page. Where necessary, clinical applications are included in italics and there are a number of clinical illustrations. We thus hope that this book will be helpful in revising and consolidating the knowledge that has been gained from the dissecting room and from more detailed and explanatory textbooks.

The anatomical drawings are the work of Jane Fallows, with help from Roger Hulley, who has transformed our rough sketches into the finished pages of illustrations that form such an important part of the book, and we should like to thank her for her patience and skill in carrying out this onerous task. Some of the drawings have been borrowed or adapted from Professor Harold Ellis's superb book *Clinical Anatomy* (9th edition), and we are most grateful to him for his permission to do this. We should also like to thank Dr Mike Benjamin of Cardiff University for the surface anatomy photographs. Finally, it is a pleasure to thank all the staff at Blackwell Science who have had a hand in the preparation of this book, particularly Fiona Goodgame and Jonathan Rowley.

Omar Faiz
David Moffat

Preface to the second edition

The preparation of the second edition has involved a thorough review of the whole text with revision where necessary. A great deal more clinical material has been added and this has been removed from the body of the text and placed at the end of each chapter as 'Clinical Notes'. In addition, four new chapters have been added containing some basic embryology, with particular reference to the clinical significance of errors of development. It is hoped that this short book will continue to offer a means of rapid revision of fundamental anatomy for both undergraduates and graduates working for the MRCS examination.

Once again, it is a pleasure to thank Jane Fallows, who prepared the illustrations for the new chapters, and all the staff at Blackwell Publishing, especially Fiona Pattison, Helen Harvey and Martin Sugden, for their help and cooperation in producing this second edition.

Omar Faiz
David Moffat

Preface to the third edition

For this third edition, the whole text and the illustrations have been reviewed and modified where necessary and two new chapters have been added on, respectively, anatomical terminology and the early development of the human embryo. In addition, a number of new illustrations have been added featuring modern imaging techniques. We hope that this book will continue to serve its purpose as a guide to 'no frills' clinical anatomy for both undergraduates and for those studying for higher degrees and diplomas.

Once again, it is a pleasure to thank the staff of Blackwell Publishing for their expert help in preparing this edition for publication, especially

Martin Davies, Jennifer Seward and Cathryn Gates. Finally, we would like to thank Jane Fallows, our artist who has been responsible for all the illustrations, old and new, that form such an important part of this book.

Omar Faiz
Simon Blackburn
David Moffat

1 Anatomical terms

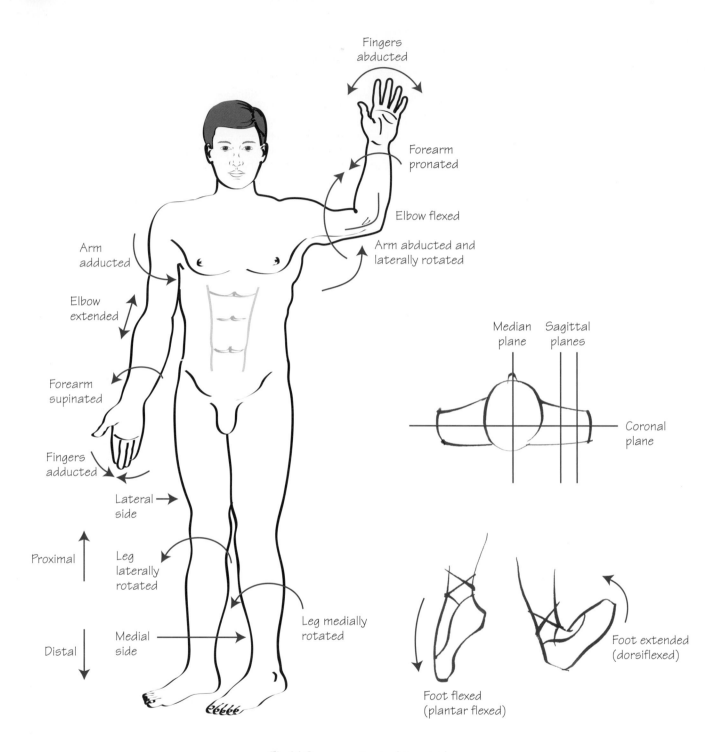

Fig.1.1 Some anatomical terminology

Correct use of anatomical terms is essential to accurate description. These terms are also essential in clinical practice to allow effective communication.

Anatomical position

It is important to appreciate that the surfaces of the body, and relative positions of structures, are described, assuming that the body is in the 'anatomical position'. In this position, the subject is standing upright with the arms by the side with the palms of the hands facing forwards. In the male the tip of the penis is pointing towards the head.

Surfaces and relative positions

- *Anterior/posterior:* the anterior surface of the body is the front, with the body in the anatomical position. The shin, for example, is referred to as the anterior aspect of the leg, regardless of its position in space. The term 'posterior' refers to the back of the body. These terms can also be used to describe relative positions. The bladder, for example, may be described as being anterior to the rectum, or the rectum posterior to the bladder.
- *Superior/inferior:* these terms refer to vertical relationships in the long axis of the body, between the head and the feet. Superior refers to the head end of the body, inferior to the foot end. These terms are most commonly used to describe relative position. The head, for example, may be described as superior to the neck. It is important to remember that the anatomical position refers to a standing subject. When a patient is lying down, their head remains superior to their neck.
- *Medial/lateral:* these terms refer to relationships relative to the midline of the body. A structure which is medial is nearer the midline, and a lateral structure is further away. So, for example, the inner thigh may be referred to as the medial part of the thigh, and the outer thigh as the lateral part. These terms are also used to describe relationships; the lung may be described as lateral to the heart, or the heart may be described as medial to the lung. In some parts of the body, these terms may cause confusion. The mobility of the forearm in space means that it is easy to get confused about which side is medial or lateral. The terms 'radial' and 'ulnar', referring to the relationship of the forearm bones, are often used instead.
- *Proximal and distal:* these terms are used to refer to relationships of structures relative to the middle of the body, the point of origin of a limb or the attachment of a muscle. These terms are commonly used to describe relationships along the length of a limb. A proximal structure is nearer the origin and a distal one further away. The hand is distal to the elbow, for example, and the elbow proximal to the hand.
- *Ventral/dorsal:* these terms are slightly different from anterior/posterior as they refer to the front and back of the body in terms of embryological development rather than the anatomical position. For the majority of the body, the anterior surface corresponds to the ventral surface and the posterior surface to the dorsal surface. The lower limb is one exception as it rotates during development such that the ventral parts come to lie posteriorly. The ventral surface of the foot, therefore, is the sole.

The ventral surface of the hand is often referred to as the palmar surface and that of the foot as the plantar surface.
- *Cranial/caudal:* These terms also refer to embryonic development. Cranial refers to the head end of the embryo, and caudal to the tail end.

Planes

Anatomical planes are used to describe sections through the body as if cut all the way through. These planes are essential to understanding cross-sectional imaging:
- *Sagittal:* this plane lies front to back, such that a sagittal section in the midline would divide the body in half through the nose and the back of the head, continuing downwards.
- *Coronal:* this plane lies at right angles to the sagittal plane and is parallel to the anterior and posterior surfaces of the body.
- *Transverse:* this plane lies across the body and is sometimes also referred to as the axial or horizontal plane. A transverse section divides the body across the middle, much like the magician sawing his assistant in half.

Movements

The following anatomical terms are used to describe movement:
- *Flexion:* is usually taken to mean the bending of a joint, such as bending the elbow or knee. Strictly, it refers to the apposition of two ventral surfaces, which is generally taken to mean the same thing.
- *Extension:* is the straightening of a joint or the movement of two ventral surfaces such that they come to lie further apart.
- *Abduction:* is movement of a part of a body away from the midline in the coronal plane. For example, abduction of the arm is lifting the arm out sideways.

In the hand, the midline is considered to be along the middle finger. Thus, abduction of the fingers refers to the motion of spreading them out. In the foot, the axis of abduction is the second toe.

The thumb is a special case. Abduction of the thumb refers to anterior movement away from the palm (see Fig. 1.1). Adduction is the opposite of this movement.
- *Adduction:* is movement towards the middle of the body in the coronal plane.
- *Plantar/dorsiflexion:* are used to describe movement of the foot at the ankle as the use of the terms 'flexion' and 'extension' is confusing. True flexion of the foot is straightening at the ankle, because this leads to two ventral surfaces coming closer together. This is, however, somewhat confusing. For this reason, the term 'plantar/flexion' is used to refer to the action of pointing the toes and dorsiflexion to refer to bending at the ankle such that the toes move towards the face.
- *Rotation:* rotation is movement around the long axis of a bone. For example rotation of the femur at the hip joint will cause the foot to point laterally or medially.
- *Supination/pronation:* are special terms used to refer to rotational movements of the forearm, best thought of when the elbow is flexed to 90 degrees. Supination refers to rotation of the forearm at the elbow laterally, such that the palm faces superiorly. Pronation refers to an inward rotation, such that the dorsal surface of the hand is uppermost.

2 Embryology

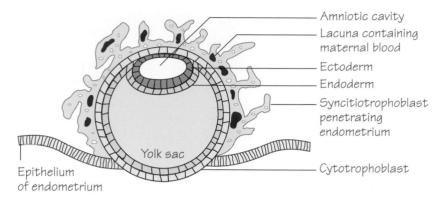

- Amniotic cavity
- Lacuna containing maternal blood
- Ectoderm
- Endoderm
- Syncitiotrophoblast penetrating endometrium
- Cytotrophoblast
- Yolk sac
- Epithelium of endometrium

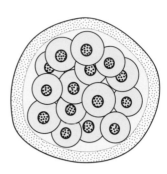

Fig.2.1
A morula, enclosed with the zona pellucida which prevents the entry of more than one spermatozoon

Fig.2.3
An almost completely implanted conceptus. The trophoblast has differentiated into the cytotrophoblast and the syncitiotrophoblast. The latter is invasive and breaks down the maternal tissue

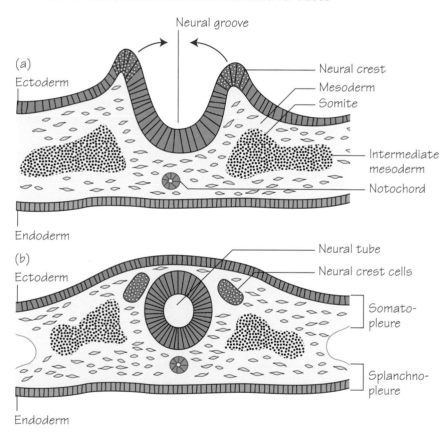

- Neural groove
- (a)
- Ectoderm
- Neural crest
- Mesoderm
- Somite
- Intermediate mesoderm
- Notochord
- Endoderm
- (b)
- Ectoderm
- Neural tube
- Neural crest cells
- Somato-pleure
- Splanchno-pleure
- Endoderm

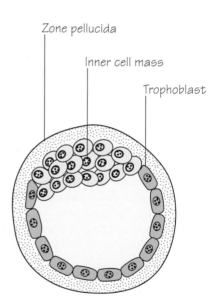

- Zone pellucida
- Inner cell mass
- Trophoblast

Fig.2.2
A blastocyst, still within the zona pellucida

Fig.2.4a, b
Two stages in the development of the neural tube. In (b) the lateral mesoderm is splitting into two layers. One layer, together with the ectoderm, forms the somatopleure and the other, together with the endoderm, forms the splanchnopleure

Normal pregnancy lasts 40 weeks. The first 8 weeks are termed the *embryonic period*, during which the body structures and organs are formed and differentiated. The *fetal period* runs from eight weeks to birth and involves growth and maturation of these structures.

The combination of ovum and sperm at fertilisation produces a *zygote*. This structure further divides to produce a ball of cells called the *morula* (Fig. 2.1), which develops into the *blastocyst* during the 4th and 5th days of pregnancy.

The blastocyst (Fig. 2.2): consists of an outer layer of cells called the *trophoblast* which encircles a fluid filled cavity. The trophoblast eventually forms the placenta. A ball of cells called the *inner cell mass* is attached to the inner surface of the trophoblast and will eventually form the embryo itself. At about six days of gestation, the blastocyst begins the process of implanting into the uterine wall. This process is complete by day 10.

Further division of the inner cell mass during the second week of development causes a further cavity to appear, the *amniotic cavity*. The blastocyst now consists of two cavities, the amniotic cavity and the *yolk sac* (derived from the original blastocyst cavity) (Fig. 2.3). These cavities are separated by the *embryonic plate*. The embryonic plate consists of two layers of cells, the *ectoderm* lying in the floor of the amniotic cavity and the *endoderm* lying in the roof of the yolk sac.

Gastrulation: is the process during the third week of gestation during which the two layers of embryonic plate divide into three, giving rise to a *trilaminar disc*. This is achieved by the development of the *primitive streak* as a thickening of the ectoderm. Cells derived from the primitive streak invaginate and migrate between the ectoderm and endoderm to form the *mesoderm*. The embryonic plate now consists of three layers:

Ectoderm: eventually gives rise to the epidermis, nervous system, anterior pituitary gland, the inner ear and the enamel of the teeth.

Endoderm: gives rise the epithelial lining of the respiratory and gastrointestinal tracts.

Mesoderm: lies between the ectoderm and endoderm and gives rise to the smooth and striated muscle of the body, connective tissue, blood vessels, bone marrow and blood cells, the skeleton, reproductive organs and the urinary tract.

The notochord and neural plate

The notochord develops from a group of ectodermal cells in the midline and eventually forms a tubular structure within the mesodermal layer of the embryo. The notochord induces development of the *neural plate* in the overlying ectoderm and eventually disappears, persisting only in the intervertebral discs as the *nucleus pulposus*.

The neural plate invaginates centrally to form a groove and then folds to form a tube by the end of week three, a process known as *neurulation* (Fig. 2.4). The neural tube then becomes incorporated into the embryo, such that it comes to lie deep to the overlying ectoderm. The resultant neural tube develops into the brain and spinal cord.

Some cells from the edge of the neural plate become separated and come to lie above and lateral to the neural tube, when they become known as *neural crest cells*. These important cells give rise to several structures including the dorsal root ganglia of spine nerves, the ganglia of the autonomic nervous system, Schwann cells, meninges, the chromaffin cells of the adrenal medulla, parafollicular cells of the thyroid and the bones of the skull and face.

Mesoderm

The mesodermal layer of the embryo comes to lie alongside the notochord and neural tube and is subdivided into three parts:

Paraxial mesoderm: lies nearest the midline and becomes segmented into paired clumps of cells called *somites*. The somites are further divided into the *sclerotome,* which eventually surrounds the neural tube and notochord to produce the vertebral column and ribs, and the *dermatomyotome* which forms the muscles of the body wall and the dermis of the skin. The segmental arrangement of the somites explains the eventual arrangement of dermatomes in the body wall and limbs (Fig. 78.1).

Intermediate mesoderm: lies lateral to the paraxial mesoderm. It eventually gives rise to the precursors of the urinary tract (see Chapter 31).

Lateral mesoderm: is involved with the formation of body cavities and the folding of the embryo (Fig. 2.4b).

A separate group of cells from the primitive streak migrate around the neural plate to form the *cardiogenic mesoderm*, which eventually gives rise to the heart.

Folding of the embryo

The folding of the embryo commences at the beginning of the fourth week (Fig. 2.5). The flat embryonic disc folds as a result of faster growth of the ectoderm cranio-caudally, such that it is concave towards the yolk sac and convex towards the amnion. Lateral folding occurs around the yolk sac in the same manner.

During this process, the lateral plate mesoderm splits to create the *embryonic coelom* or body cavity (Fig. 2.4). The inner layer is called the *splanchnopleure* and surrounds the yolk sac in such a way that it becomes incorporated into the embryo, forming the cells lining the lumen of the gastrointestinal tract. The cranial part of the yolk sac migrates further cranially, forming the foregut, and the caudal part migrates further caudally, forming the hindgut (Fig. 2.6). As the folding of the embryo continues the yolk sac forms a small vesicle lying outside the embryo and connected to the gut by a narrow *vitello-intestinal duct* (see Chapter 31). The two ends of the primitive gut are separated from the amniotic cavity at the cranial end by the *buccopharyngeal membrane,* and the caudal end by the *cloacal membrane*, which are formed of ectoderm and endoderm with no intervening mesoderm. They eventually disappear to form cranial and caudal openings into the pharynx and the anal canal, respectively.

The outer layer of the lateral mesoderm is called the *somatopleure*. This layer is invaded by paraxial mesoderm, forming the body wall muscles. Outgrowths from the somatopleure form the limbs, which appear as buds during the 4th week of gestation.

At the end of the process of folding, the embryo contains a single internal cavity, the intra-embryonic coelom, which is eventually separated by the formation of the diaphragm into pleural and peritoneal cavities.

During this period of folding, the branchial arches develop and form a number of structures described in Chapter 76.

Between the 4th and 8th week of gestation, the limb buds, facial structures, palate, digits, gonads and genitalia, all start to differentiate, such that by the end of week eight all the external and internal structures required are present.

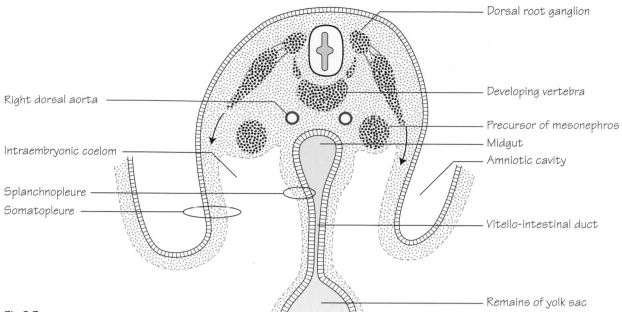

Fig.2.5
Lateral folding of the embryo so that it projects into the amniotic cavity.
Striated muscle, from the somites, is growing down into the somatopleure
(body wall) taking its nerve supply with it. Smooth muscle of the gut will
develop in the mesoderm of the splanchnopleure

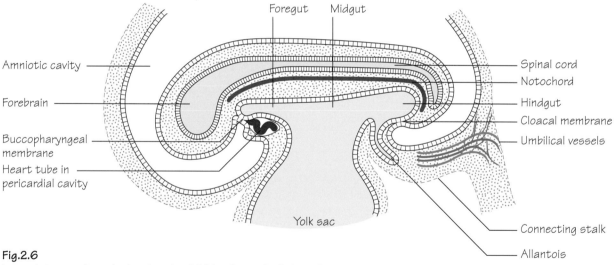

Fig.2.6
Lateral view to show the head and tail folds. The neck of the yolk sac will later
close off, leaving the midgut intact. The allantois is functionless and will later
degenerate to form the median umbilical ligament. The connecting stalk contains
the umbilical vessels (intraembryonic course not shown)

Clinical notes

Sacrococcygeal teratomas: these rare tumours arise as a result of failure of the normal obliteration of the primitive streak. As the primitive streak contains cells which are capable of producing cells from all three germ cell layers (ectoderm, mesoderm and endoderm), these tumours contain elements of tissues derived from all of them.

Neural tube defects: failure of the neural plate to completely fold to form the neural tube can cause abnormalities in the formation of the central nervous system. At the most extreme, the brain fails to develop completely (*anencephaly*). Failure of closure of the neural tube can also cause abnormalities of the overlying structures. *Spina bifida*, for example, results from failure of normal fusion of the posterior part of the vertebral column (see Chapter 77).

3 The thoracic wall I

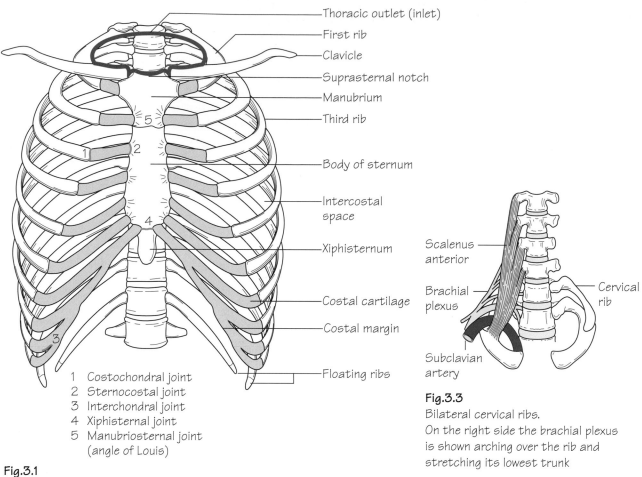

1 Costochondral joint
2 Sternocostal joint
3 Interchondral joint
4 Xiphisternal joint
5 Manubriosternal joint
   (angle of Louis)

Fig.3.1
The thoracic cage. The outlet (inlet)
of the thorax is outlined

Fig.3.3
Bilateral cervical ribs.
On the right side the brachial plexus
is shown arching over the rib and
stretching its lowest trunk

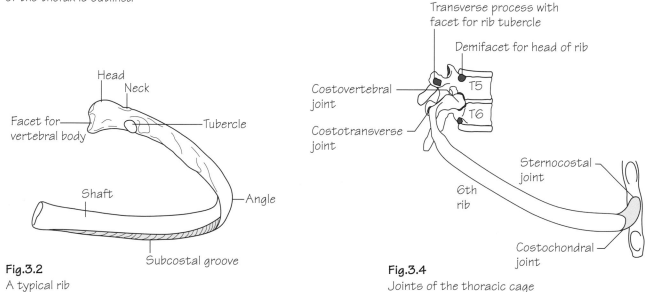

Fig.3.2
A typical rib

Fig.3.4
Joints of the thoracic cage

 Anatomy at a Glance, Third Edition. Omar Faiz, Simon Blackburn and David Moffat.
© 2011 Blackwell Publishing Ltd. Published 2011 by Blackwell Publishing Ltd.

The thoracic cage

The thoracic cage is formed by the sternum and costal cartilages in front, the vertebral column behind and the ribs and intercostal spaces laterally.

It is separated from the abdominal cavity by the diaphragm and communicates superiorly with the root of the neck through the *thoracic inlet* (Fig. 3.1).

The ribs (Fig. 3.1)

• Of the 12 pairs of ribs, the first seven articulate with the vertebrae posteriorly and with the sternum anteriorly by way of the costal cartilages (*true ribs*).
• The cartilages of the 8th, 9th and 10th ribs articulate with the cartilages of the ribs above (*false ribs*).
• The 11th and 12th ribs are termed 'floating' because they do not articulate anteriorly (*false ribs*).

Typical ribs (3rd–9th)

These comprise the following features (Fig. 3.2):
• A *head* which bears two demifacets for articulation with the bodies of the numerically corresponding vertebra and the vertebra above (Fig. 3.4).
• A *tubercle* which comprises a rough non-articulating lateral facet as well as a smooth medial facet, which articulates with the transverse process of the corresponding vertebra (Fig. 3.4).
• A *subcostal groove* which is the hollow on the inferior inner aspect of the shaft accommodating the intercostal neurovascular structures.

Atypical ribs (1st, 2nd, 10th, 11th, 12th)

• The **1st rib** (see Fig. 68.2) is short, flat and sharply curved. The head bears a single facet for articulation. A prominent tubercle (*scalene tubercle*) on the inner border of the upper surface represents the insertion site for scalenus anterior. The subclavian vein passes over the 1st rib anterior to this tubercle, whereas the subclavian artery and lowest trunk of the brachial plexus pass posteriorly.
• The **2nd rib** is less curved and longer than the 1st rib.
• The **10th rib** has only one articular facet on the head.
• The **11th** and **12th ribs** are short and do not articulate anteriorly. They articulate posteriorly with the vertebrae by way of a single facet on the head. They are devoid of both a tubercle and a subcostal groove.

The sternum (Fig. 3.1)

The sternum comprises a manubrium, body and xiphoid process.
• The *manubrium* has facets for articulation with the clavicles, 1st costal cartilage and upper part of the 2nd costal cartilage. It articulates inferiorly with the body of the sternum at the *manubriosternal joint*.
• The *body* is composed of four parts or *sternebrae* which fuse between 15 and 25 years of age. It has facets for articulation with the lower part of the 2nd and the 3rd to 7th costal cartilages.
• The *xiphoid* articulates above with the body at the *xiphisternal joint*. The xiphoid usually remains cartilaginous well into adult life.

Costal cartilages

These are bars of hyaline cartilage which connect the upper seven ribs directly to the sternum and the 8th, 9th and 10th ribs to the cartilage immediately above.

Joints of the thoracic cage
(Figs. 3.1 and 3.4)

• The *manubriosternal joint* is a symphysis (a joint in which the bone ends are covered with two layers of hyaline cartilage which are themselves joined by fibrocartilage). It usually ossifies after the age of 30 years.
• The *xiphisternal joint* is also a symphysis.
• The *1st sternocostal joint* is a primary cartilaginous joint (a joint in which the two bones are directly joined by a single layer of hyaline cartilage). The rest (2nd to 7th) are synovial joints (joints which include a cavity containing synovial fluid and lined by synovial membrane). All have a single synovial joint except for the 2nd which is double.
• The *costochondral joints* (between the ribs and costal cartilages) are primary cartilaginous joints.
• The *interchondral joints* (between the costal cartilages of the 8th, 9th and 10th ribs) are synovial joints.
• The *costovertebral joints* comprise two synovial joints formed by the articulations of the demifacets on the head of each rib with the bodies of its corresponding vertebra, together with that of the vertebra above. The 1st and 10th–12th ribs have a single synovial joint with their corresponding vertebral bodies.
• The *costotransverse joints* are synovial joints formed by the articulations between the facets on the rib tubercle and the transverse process of its corresponding vertebra.

Clinical notes

• **Cervical rib**: a cervical rib is a rare 'extra' rib which articulates with C7 posteriorly and the 1st rib anteriorly. A neurological deficit and vascular insufficiency arise as a result of pressure from the rib on the lowest trunk of the brachial plexus (T1) and subclavian artery, respectively (Fig. 3.3).
• **Rib fracture**: although significant injury is generally required to damage the bony thoracic wall, pathological rib fractures (i.e. fractures occurring in diseased bone – usually metastatic carcinoma) can result from minimal trauma. Many rib fractures are not visible on X-rays unless complications, such as a pneumothorax or a haemothorax, are present. Treatment of simple rib fractures aims to relieve pain, as inadequate analgaesia can lead to poor chest expansion and consequent pneumonia. In severe trauma, multiple rib fractures can give rise to a 'flail' segment, in which two or more ribs are fractured in two or more places. When this occurs, ventilatory compromise can supervene. This usually results from associated traumatic lung injury but is also exacerbated by paradoxical movement of the 'floating' flail segment with respiration.
• **Pectus excavatum and carinatum**: deformities of the chest wall are uncommon. Pectus excavatum represents a visible furrow in the anterior chest wall that results from a depressed sternum. In contrast, pectus carinatum (pigeon chest) is a clinical manifestation that results from a sternal protrusion. Rarely do either of these conditions require surgical correction.

4 The thoracic wall II

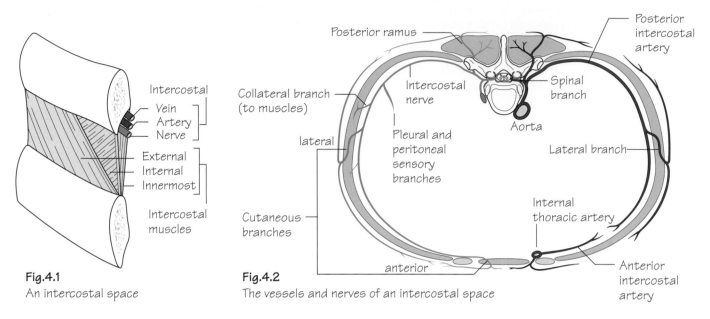

Fig.4.1
An intercostal space

Fig.4.2
The vessels and nerves of an intercostal space

Fig.4.3
The diaphragm

 Anatomy at a Glance, Third Edition. Omar Faiz, Simon Blackburn and David Moffat.

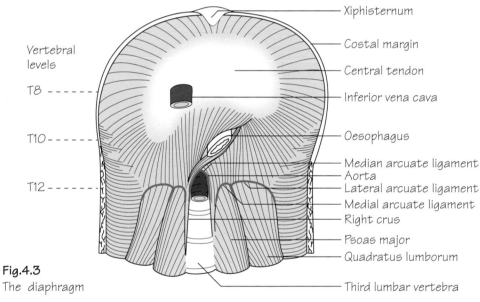

The intercostal space (Fig. 4.1)

Typically, each space contains three muscles comparable to those of the abdominal wall. These include the:

- **External intercostal**: this muscle fills the intercostal space from the vertebra posteriorly to the costochondral junction anteriorly where it becomes the thin anterior intercostal membrane. The fibres run downwards and forwards from rib above to rib below.
- **Internal intercostal**: this muscle fills the intercostal space from the sternum anteriorly to the angles of the ribs posteriorly where it becomes the posterior intercostal membrane which reaches as far back as the vertebral bodies. The fibres run downwards and backwards.
- **Innermost intercostals**: this group comprises the *subcostal* muscles posteriorly, the *intercostales intimi* laterally and the *transversus thoracis* anteriorly. The fibres of these muscles span more than one intercostal space.

The neurovascular space is the plane in which the neurovascular bundle (intercostal vein, artery and nerve) courses. It lies between the internal intercostal and innermost intercostal muscle layers.

The intercostal structures course under cover of the subcostal groove.

Vascular supply and venous drainage of the chest wall

The intercostal spaces receive their *arterial supply* from the anterior and posterior intercostal arteries.

- The *anterior intercostal arteries* are branches of the internal thoracic artery and its terminal branch, the musculophrenic artery. The lowest two spaces have no anterior intercostal supply (Fig. 4.2).
- The first 2–3 *posterior intercostal arteries* arise from the superior intercostal branch of the costocervical trunk, a branch of the 2nd part of the subclavian artery (see Fig. 65.1). The lower nine posterior intercostal arteries are branches of the thoracic aorta. The posterior intercostal arteries are much longer than the anterior intercostal arteries (Fig. 4.2).

The anterior intercostal *veins* drain anteriorly into the internal thoracic and musculophrenic veins. The posterior intercostal veins drain into the azygos and hemiazygos systems (see Fig. 6.2).

Lymphatic drainage of the chest wall

Lymph drainage from the:

- **Anterior chest wall** is to the anterior axillary nodes.
- **Posterior chest wall** is to the posterior axillary nodes.
- **Anterior intercostal spaces** is to the internal thoracic nodes.
- **Posterior intercostal spaces** is to the para-aortic nodes.

Nerve supply of the chest wall (Fig. 4.2)

The intercostal nerves are the anterior primary rami of the thoracic segmental nerves. Only the upper six intercostal nerves reach the sternum, the remainder run initially in their intercostal spaces, then within the muscles of the abdominal wall, eventually gaining access to its anterior aspect.

Branches of the intercostal nerves include:

- *Cutaneous* anterior and lateral branches.
- A *collateral* branch which supplies the muscles of the intercostal space (also supplied by the main intercostal nerve).
- *Sensory* branches from the pleura (upper nerves) and peritoneum (lower nerves).

Exceptions include:

- The 1st intercostal nerve is joined to the brachial plexus and has no anterior cutaneous branch.
- The 2nd intercostal nerve is joined to the medial cutaneous nerve of the arm by the intercostobrachial nerve branch. The 2nd intercostal nerve consequently supplies the skin of the armpit and medial side of the arm.

The diaphragm (Fig. 4.3)

The diaphragm separates the thoracic and abdominal cavities. It is composed of a peripheral muscular portion which inserts into a central aponeurosis—the *central tendon*.

The muscular part has three component origins:

- A *vertebral part* which comprises the crura and arcuate ligaments. The right crus arises from the front of the L1–3 vertebral bodies and intervening discs. Some fibres from the right crus pass around the lower oesophagus.

The left crus originates from L1 and L2 only.

The medial arcuate ligament is made up of thickened fascia which overlies psoas major and is attached medially to the body of L1 and laterally to the transverse process of L1. The lateral arcuate ligament is made up of fascia which overlies quadratus lumborum from the transverse process of L1 medially to the 12th rib laterally.

The median arcuate ligament is a fibrous arch which connects left and right crura.

- A *costal part* attached to the inner aspects of the lower six ribs.
- A *sternal part* which consists of two small slips arising from the deep surface of the xiphoid process.

Openings in the diaphragm

Structures traverse the diaphragm at different levels to pass from thoracic to abdominal cavities and vice versa. These levels are as follows:

- T8, the *opening for the inferior vena cava*: transmits the inferior vena cava and right phrenic nerve.
- T10, the *oesophageal opening*: transmits the oesophagus, vagi and branches of the left gastric artery and vein.
- T12, the *aortic opening*: transmits the aorta, thoracic duct and azygos vein.

The left phrenic nerve passes into the diaphragm as a solitary structure, having passed down the left side of the pericardium (Fig. 9.1).

Nerve supply of the diaphragm

- **Motor supply**: the entire motor supply arises from the phrenic nerves (C3,4,5). Diaphragmatic contraction is the mainstay of inspiration.
- **Sensory supply**: the periphery of the diaphragm receives sensory fibres from the lower intercostal nerves. The sensory supply from the central part is carried by the phrenic nerves.

Clinical notes

- **Diaphragmatic herniae**: the diaphragm is formed by the embryological fusion of the septum transversum, dorsal mesentery and pleuro-peritoneal membranes. Failed fusion results in congenital diaphragmatic herniae. Most commonly, congenital herniation occurs through the Bochdalek foramen posteriorly (through the pleuroperitoneal canal), it may also occur through the Morgagni foramen anteriorly (between the xiphoid, costal cartilages and the attached diaphragm). Acquired diaphragmatic hernia occurs frequently. The most common type of this kind is the hiatus hernia. It represents a weakening of the oesophageal hiatus. This condition occurs mostly in adulthood and often gives rise to symptomatic acid reflux. The majority of patients require medical treatment only, but some require surgical correction.

The mediastinum I – the contents of the mediastinum

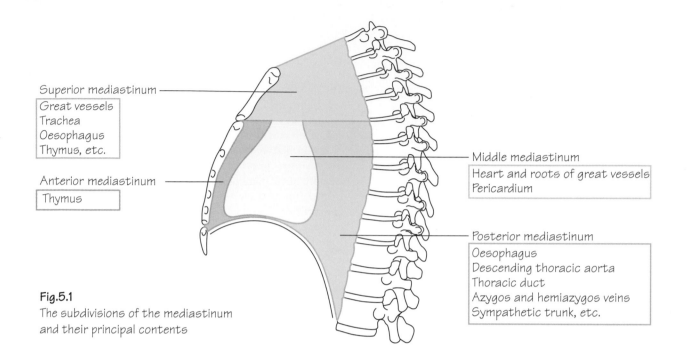

Superior mediastinum

Great vessels
Trachea
Oesophagus
Thymus, etc.

Anterior mediastinum

Thymus

Middle mediastinum

Heart and roots of great vessels
Pericardium

Posterior mediastinum

Oesophagus
Descending thoracic aorta
Thoracic duct
Azygos and hemiazygos veins
Sympathetic trunk, etc.

Fig.5.1

The subdivisions of the mediastinum
and their principal contents

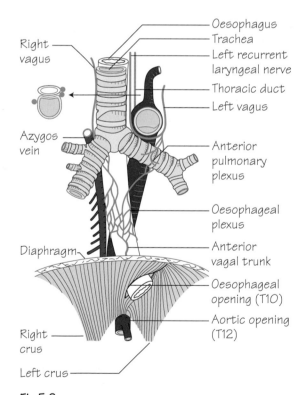

Right vagus

Oesophagus
Trachea
Left recurrent laryngeal nerve
Thoracic duct
Left vagus

Azygos vein

Anterior pulmonary plexus

Oesophageal plexus

Anterior vagal trunk

Diaphragm

Oesophageal opening (T10)

Aortic opening (T12)

Right crus

Left crus

Fig.5.2

The course and principal relations of the oesophagus.
Note that it passes through the *right crus* of the
diaphragm

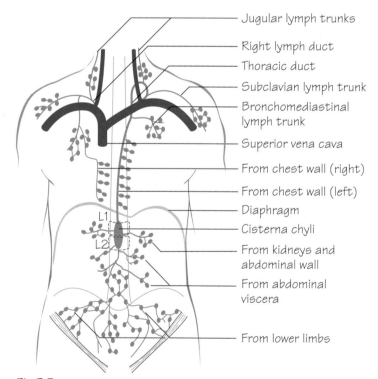

Jugular lymph trunks

Right lymph duct

Thoracic duct

Subclavian lymph trunk

Bronchomediastinal lymph trunk

Superior vena cava

From chest wall (right)

From chest wall (left)

Diaphragm

Cisterna chyli

From kidneys and abdominal wall

From abdominal viscera

From lower limbs

L1

L2

Fig.5.3

The thoracic duct and its areas of drainage.
The right lymph duct is also shown

Anatomy at a Glance, Third Edition. Omar Faiz, Simon Blackburn and David Moffat.
© 2011 Blackwell Publishing Ltd. Published 2011 by Blackwell Publishing Ltd.

Subdivisions of the mediastinum
(Fig. 5.1)

The mediastinum is the space located between the two pleural sacs. For descriptive purposes, it is divided into superior and inferior mediastinal regions by a line drawn backwards horizontally from the angle of Louis (manubriosternal joint) to the vertebral column (T4/5 intervertebral disc).

The *superior mediastinum* communicates with the root of the neck through the 'superior thoracic aperture' (thoracic inlet). The latter opening is bounded anteriorly by the manubrium, posteriorly by T1 vertebra and laterally by the 1st rib.

The *inferior mediastinum* is further subdivided into the:

- **Anterior mediastinum** which is the region in front of the pericardium.
- **Middle mediastinum** which consists of the pericardium and heart.
- **Posterior mediastinum** which is the region between the pericardium and vertebrae.

The contents of the mediastinum
(Figs. 5.1, 5.2, and 8.2)

The oesophagus

- **Course**: the oesophagus commences as a cervical structure at the level of the cricoid cartilage at C6 in the neck. In the thorax, the oesophagus passes initially through the superior and then the posterior mediastina. Having deviated slightly to the left in the neck, the oesophagus returns to the midline in the thorax at the level of T5. From here, it passes downwards and forwards to reach the oesophageal opening in the diaphragm (T10).
- **Structure**: the oesophagus is composed of four layers:
  - An inner mucosa of stratified squamous epithelium.
  - A submucous layer.
  - A double muscular layer – longitudinal outer layer and circular inner layer. The muscle is striated in the upper two-thirds and smooth in the lower third.
  - An outer layer of areolar tissue.
- **Relations**: the lateral relations of the oesophagus are shown in Fig. 5.2. On the right side, the oesophagus is crossed only by the azygos vein and the right vagus nerve, which, therefore, represents the least hazardous surgical approach. Anteriorly, the oesophagus is related to the trachea and left bronchus in the upper thorax and the pericardium overlying the left atrium in the lower thorax. Posterior relations of the oesophagus include the thoracic vertebrae, the thoracic duct and azygos veins. In the lower thorax, the aorta is a posterior oesophageal relation.
- **Arterial supply and venous drainage**: Owing to its length (25 cm), the oesophagus receives arterial blood from different sources throughout its course:
  - *Upper third*: inferior thyroid artery.
  - *Middle third*: oesophageal branches of thoracic aorta.
  - *Lower third*: left gastric branch of coeliac artery.
  The venous drainage is similarly varied throughout its length:
  - *Upper third*: inferior thyroid veins.
  - *Middle third*: azygos system.
  - *Lower third*: both the azygos (systemic system) and left gastric (portal system) veins.
- **Lymphatic drainage**: is to a peri-oesophageal lymph plexus and then to the posterior mediastinal nodes. From here, lymph drains into supraclavicular nodes. The lower oesophagus also drains into the nodes around the left gastric vessels.

The thoracic duct (Fig. 5.3)

- The *cisterna chyli* is a lymphatic sac that receives lymph from the abdomen and lower half of the body. It is situated between the abdominal aorta and the right crus of the diaphragm.
- The *thoracic duct* carries lymph from the cisterna chyli through the thorax to drain into the left brachiocephalic vein. It usually receives tributaries from the left jugular, subclavian and mediastinal lymph trunks, although these may open into the large neck veins directly.
- On the right side, the main lymph trunks from the right upper body usually join and drain directly through a common tributary, the *right lymph duct*, into the right brachiocephalic vein.

The thymus gland

- This is an important component of the lymphatic system. It usually lies behind the manubrium (in the superior mediastinum), but can extend to about the 4th costal cartilage in the anterior mediastinum. After puberty the thymus is gradually replaced by fat.

Clinical notes

- **Oesophageal varices**: the dual portal and systemic venous drainage of the lower third of the oesophagus forms a site of porto-systemic anastomosis (a site at which veins draining into the portal circulation and those draining into the systemic circulation are in continuity). In advanced liver cirrhosis, portal pressure rises, resulting in back-pressure on the left gastric tributaries at the lower oesophagus, causing these veins to become distended and fragile (oesophageal varices). This predisposes them to rupture, which causes potentially life-threatening haemorrhage.
- **Oesophageal carcinoma**: carries an extremely poor prognosis. Two main histological types, squamous and adenocarcinoma, account for the majority of tumours. The incidence of adenocarcinoma of the lower third of the oesophagus is currently increasing for unknown reasons. Most tumours are unresectable at the time of diagnosis. The insertion of stents and the use of lasers to pass through tumour obstruction have become the principal methods of palliation. Where oesophageal tumour resection is possible, the approach varies depending on the location of the tumour. The options include a left thoraco-abdominal approach or a two-stage 'Ivor-Lewis' approach (a right thoracotomy and laparotomy) for low oesophageal lesions. In contrast, for high oesophageal lesions, a three-stage 'McKeown' oesophagectomy (a cervical incision, right thoracotomy and laparotomy) or transhiatal oesophagectomy is required.
- **Oesophagogastroduodenoscopy (OGD)**: is usually performed under sedation with a flexible fibre-optic endoscope. This technique is used to visualise the oesophageal mucosa, but also permits biopsies to be taken. In an adult, the endoscope will require insertion to 15 cm to reach the cricopharyngeal constriction (a narrowing of the oesophgus at the level of the cricopharyngeus muscle), to 25 cm to reach the level of the aortic arch as it passes over the left main bronchus and to 40 cm to reach the squamocolumnar junction, where the oesophageal mucosa meets the gastric mucosa. Beyond this point, the endoscope passes into the stomach.

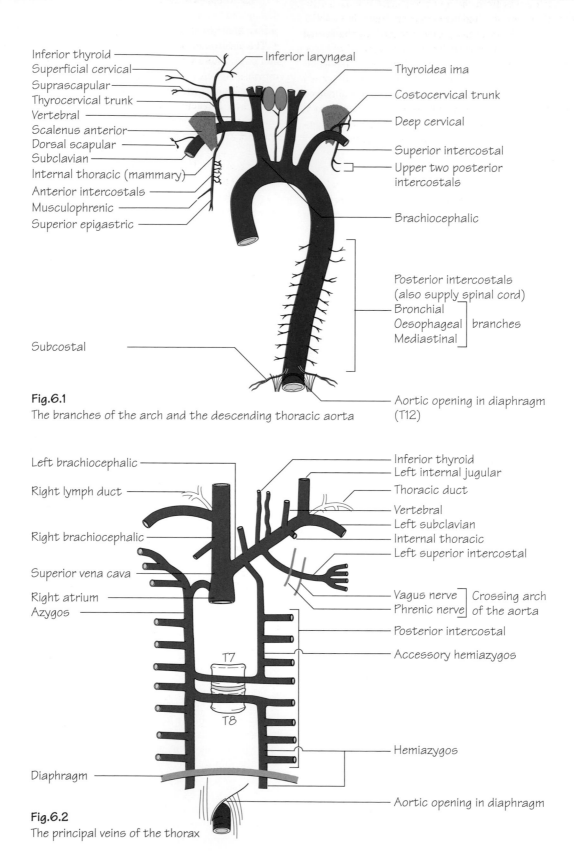

Inferior thyroid — Inferior laryngeal
Superficial cervical — Thyroidea ima
Suprascapular — Costocervical trunk
Thyrocervical trunk — Deep cervical
Vertebral — Superior intercostal
Scalenus anterior — Upper two posterior intercostals
Dorsal scapular —
Subclavian —
Internal thoracic (mammary) — Brachiocephalic
Anterior intercostals —
Musculophrenic — Posterior intercostals (also supply spinal cord)
Superior epigastric — Bronchial
Oesophageal } branches
Mediastinal
Subcostal — Aortic opening in diaphragm (T12)

Fig.6.1
The branches of the arch and the descending thoracic aorta

Left brachiocephalic — Inferior thyroid
— Left internal jugular
Right lymph duct — Thoracic duct
— Vertebral
— Left subclavian
Right brachiocephalic — Internal thoracic
— Left superior intercostal
Superior vena cava —
Right atrium — Vagus nerve ⎤ Crossing arch
Azygos — Phrenic nerve ⎦ of the aorta
— Posterior intercostal
— Accessory hemiazygos
T7
T8
— Hemiazygos
Diaphragm —
— Aortic opening in diaphragm

Fig.6.2
The principal veins of the thorax

The thoracic aorta (Fig. 6.1)

The *ascending aorta* arises from the aortic vestibule behind the infundibulum of the right ventricle and the pulmonary trunk. It is continuous with the *aortic arch*. The arch lies posterior to the lower half of the manubrium and arches from front to back over the left main bronchus. The *descending thoracic aorta* is continuous with the arch and begins at the lower border of the body of T4. It initially lies slightly to the left of the midline and then passes medially to gain access to the abdomen by passing beneath the median arcuate ligament of the diaphragm at the level of T12. From here, it continues as the abdominal aorta.

The branches of the ascending aorta are the *right* and *left coronary arteries*.

- The branches of the aortic arch are the:
  - *Brachiocephalic artery*: arises from the arch behind the manubrium and courses upwards to bifurcate into *right subclavian* and *right common carotid branches* posterior to the right sternoclavicular joint.
  - *Left common carotid artery*: see p. 147.
  - *Left subclavian artery*.
  - *Thyroidea ima artery*.
  - The branches of the descending thoracic aorta include the *oesophageal, bronchial, mediastinal, posterior intercostal* and *subcostal arteries*.

The subclavian arteries (see Fig. 65.1)

The subclavian arteries become the axillary arteries at the outer border of the 1st rib. Each artery is divided into three parts by scalenus anterior:

- **1st part**: the part of the artery that lies medial to the medial border of scalenus anterior. It gives rise to three branches: the *vertebral artery* (p. 149), *thyrocervical trunk* and *internal thoracic (mammary) artery*. The latter artery courses on the posterior surface of the anterior chest wall, one finger's breadth from the lateral border of the sternum. Along its course, it gives off anterior intercostal, thymic and perforating branches. The 'perforators' pass through the anterior chest wall to supply the breast. The internal thoracic artery divides behind the 6th costal cartilage into superior epigastric and musculophrenic branches. The thyrocervical trunk terminates as the inferior thyroid artery.
- **2nd part**: the part of the artery that lies behind scalenus anterior. It gives rise to the *costocervical trunk* (see Fig. 65.1).
- **3rd part**: the part of the artery that lies lateral to the lateral border of scalenus anterior. This part gives rise to the *dorsal scapular artery*.

The great veins (Fig. 6.2)

The *brachiocephalic veins* are formed by the confluence of the *subclavian* and *internal jugular* veins behind the sternoclavicular joints. The left brachiocephalic vein traverses diagonally behind the manubrium to join the right brachiocephalic vein behind the 1st costal cartilage, thus forming the *superior vena cava*. The superior vena cava receives only one tributary – the *azygos vein*.

The azygos system of veins (Fig. 6.2)

- The *azygos vein*: commences as the union of the right subcostal vein and one or more veins from the abdomen. It passes through the aortic opening in the diaphragm, ascends on the posterior chest wall to the level of T4 and then arches over the right lung root to enter the superior vena cava. It receives tributaries from the lower eight right posterior intercostal veins, right superior intercostal vein and hemiazygos and accessory hemiazygos veins.
- The *hemiazygos vein*: arises on the left side in the same manner as the azygos vein. It passes through the aortic opening in the diaphragm and up to the level of T9, from where it passes diagonally behind the aorta and thoracic duct to drain into the azygos vein at the level of T8. It receives venous blood from the lower four left posterior intercostal veins.
- The *accessory hemiazygos vein*: drains blood from the middle posterior intercostal veins (as well as some bronchial and mid-oesophageal veins). The accessory hemiazygos crosses to the right to drain into the azygos vein at the level of T7.
- The upper four left intercostal veins drain into the left brachiocephalic vein via the left superior intercostal vein.

Clinical notes

- **Aortic dissection**: the majority of dissections commence in the ascending aorta. Severely hypertensive patients, as well as those with Marfan's syndrome, are most at risk of developing this condition. Aortic dissection can also occur secondary to chest trauma. Dissection arises when the aortic intima is torn, allowing blood to track between the layers of the aortic wall, thereby compromising the blood flow to significant vessels. A dissection will usually extend distally to involve the arteries of the head and neck and, ultimately, the renal, spinal and iliac arteries when the abdominal aorta is reached. Proximal extension to the aortic root may also occur, leading to aortic regurgitation. The sudden onset of severe central chest pain radiating to the back suggests dissection, but myocardial infarction requires exclusion. A widened mediastinum is sometimes visible on X-ray, but CT scanning is diagnostic. Treatment relies on hypertension control and surgery.
- **Subclavian steal syndrome**: this condition occurs infrequently. It arises as a result of obstruction to blood flow in the first part of the subclavian artery. In consequence, the vertebral artery provides a collateral supply to the arm by reversing its flow and thereby depleting the cerebral circulation. Classical symptoms include syncope and visual disturbance on exercising the arm with the compromised blood supply.

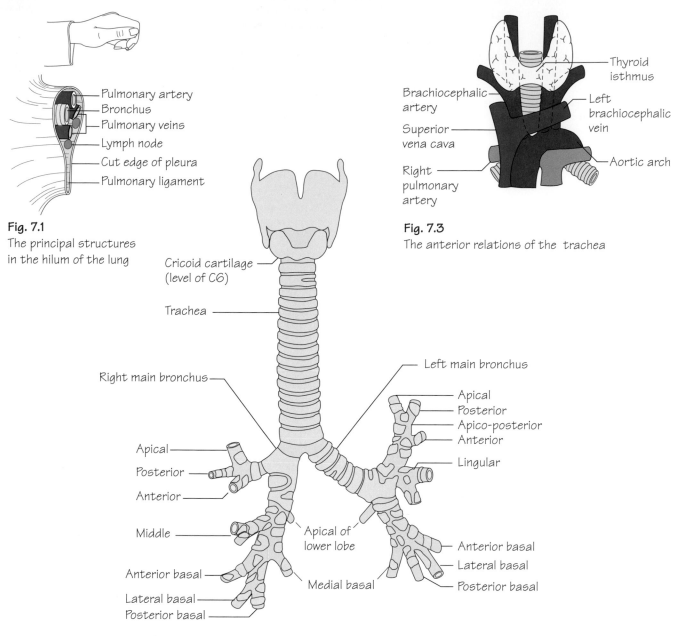

Fig. 7.1
The principal structures
in the hilm of the lung

- Pulmonary artery
- Bronchus
- Pulmonary veins
- Lymph node
- Cut edge of pleura
- Pulmonary ligament

Fig. 7.3
The anterior relations of the trachea

- Thyroid isthmus
- Brachiocephalic artery
- Left brachiocephalic vein
- Superior vena cava
- Aortic arch
- Right pulmonary artery

Cricoid cartilage (level of C6)

Trachea

Right main bronchus

Left main bronchus

Apical
Posterior
Anterior

Apical
Posterior
Apico-posterior
Anterior
Lingular

Middle

Apical of lower lobe

Anterior basal
Lateral basal

Anterior basal
Lateral basal
Posterior basal

Medial basal

Posterior basal

Fig. 7.2
The trachea and main bronchi

The respiratory tract is separated into upper and lower parts for the purposes of description. The upper respiratory tract comprises the nasopharynx and larynx, whereas the lower is comprised of the trachea, bronchi and lungs.

The pleurae
• Each pleura consists of two layers: a *visceral layer* which is adherent to the lung and a *parietal layer* which lines the inner aspect of the chest wall, diaphragm and sides of the pericardium and mediastinum.
• At the hilum of the lung the visceral and parietal layers become continuous. This cuff hangs loosely over the hilum and is known as the *pulmonary ligament*. It permits expansion of the pulmonary veins and movement of hilar structures during respiration (Fig. 7.1).
• The two pleural cavities do not connect.
• The pleural cavity contains a small amount of pleural fluid which acts as a lubricant, decreasing friction between the pleurae.
• During maximal inspiration, the lungs almost fill the pleural cavities. In quiet inspiration, the lungs do not expand fully into the costodiaphragmatic and costomediastinal recesses of the pleural cavity.
• The parietal pleura is sensitive to pain and touch (carried by the somatic intercostal and phrenic nerves). The visceral pleura is sensitive only to stretch (carried by autonomic afferents from the pulmonary plexus).

The trachea (Fig. 7.2)
• **Course**: the trachea commences at the level of the cricoid cartilage in the neck (C6). It terminates at the level of the manubriosternal joint, or angle of Louis (T4/5) where it bifurcates into right and left main bronchi.
• **Structure**: the trachea is a rigid fibro-elastic structure. Incomplete rings of hyaline cartilage continuously maintain the patency of the lumen. The trachea is lined internally with ciliated columnar epithelium.
• **Relations**: the oesophagus lies posterior to the trachea throughout its length. The 2nd, 3rd and 4th tracheal rings are crossed anteriorly by the thyroid isthmus (Figs. 7.3 and 69.1).
• **Blood supply**: the trachea receives its blood supply from branches of the inferior thyroid and bronchial arteries.

The bronchi and bronchopulmonary segments (Fig. 7.2)
• The right main bronchus is shorter, wider and takes a more vertical course than the left. The width and vertical course of the right main bronchus account for the tendency for inhaled foreign bodies to preferentially impact in the right middle and lower lobe bronchi.

• The left main bronchus enters the hilum and divides into superior and inferior lobar bronchi. The right main bronchus gives off the bronchus to the upper lobe prior to entering the hilum and, once into the hilum, divides into middle and inferior lobar bronchi.
• Each lobar bronchus divides within the lobe into segmental bronchi. Each segmental bronchus enters a bronchopulmonary segment.
• Each bronchopulmonary segment is pyramidal in shape with its apex directed towards the hilum (Fig. 8.1). It is a structural unit of a lobe that has its own segmental bronchus, artery and lymphatics. If one bronchopulmonary segment is diseased, it may, therefore, be resected with preservation of the rest of the lobe. The veins draining each segment are intersegmental.

Clinical notes

• **Pneumothorax**: air can enter the pleural cavity following a fractured rib, causing a minor lung tear. This eliminates the normal negative pleural pressure, causing the lung to collapse. Significant pneumothoraces require the insertion of a chest drain into the pleural cavity. The presence of a chest drain with an underwater seal, which permits air to flow out of the chest but not back in, allows drainage of the pleural air and expansion of the lung. If the pleural tear acts as a one-way flap-valve, air can enter but not exit the pleural cavity. This results in a tension pneumothorax. This is a medical emergency, as failure to relieve the pneumothorax results in mediastinal shift to the contralateral side, causing cardiovascular compromise and eventual cardiac arrest.
• **Pleurisy**: inflammation of the pleura (pleurisy) results from infection of the adjacent lung (pneumonia). When this occurs, the inflammatory process renders the pleura sticky. Under these circumstances, a pleural rub can often be auscultated over the affected region during inspiration and expiration. Pus in the pleural cavity (secondary to an infective process) is termed an empyema. The latter often results in significant systemic toxicity and requires pleural drainage.
• **Bronchial carcinoma**: is the most common cancer amongst men in the United Kingdom. Four main histological types occur, with small cell carcinoma carrying the worst prognosis. The overall prognosis remains appalling, with only 10% of sufferers surviving for 5 years. It occurs most commonly in the mucous membranes lining the major bronchi near the hilum. Local invasion and spread to hilar and tracheobronchial nodes occur early.

8 The lungs

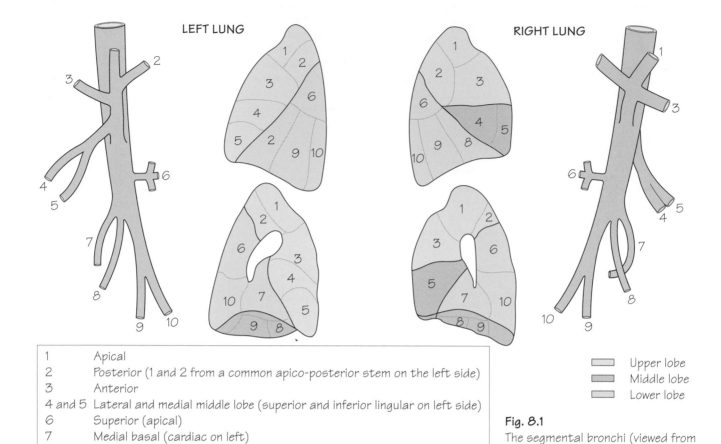

1	Apical
2	Posterior (1 and 2 from a common apico-posterior stem on the left side)
3	Anterior
4 and 5	Lateral and medial middle lobe (superior and inferior lingular on left side)
6	Superior (apical)
7	Medial basal (cardiac on left)
8	Anterior basal (7 and 8 often by a common stem on left)
9	Lateral basal
10	Posterior basal

Upper lobe
Middle lobe
Lower lobe

Fig. 8.1
The segmental bronchi (viewed from the lateral side) and the broncho-pulmonary segments, with their standard numbering

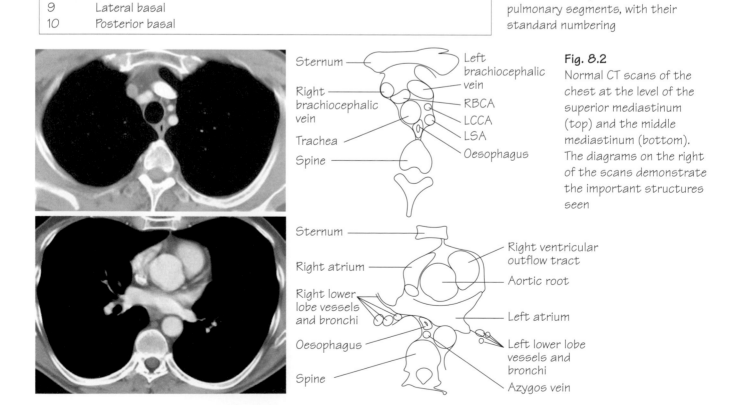

Fig. 8.2
Normal CT scans of the chest at the level of the superior mediastinum (top) and the middle mediastinum (bottom). The diagrams on the right of the scans demonstrate the important structures seen

Anatomy at a Glance, Third Edition. Omar Faiz, Simon Blackburn and David Moffat.

The lungs (Fig. 8.1)

- The lungs provide an alveolar surface area of approximately 40 m^2 for gaseous exchange.
- Each lung has: an *apex* which reaches above the sternal end of the 1st rib, a *costovertebral* surface which underlies the chest wall, a *base* overlying the diaphragm and a *mediastinal* surface which is moulded to adjacent mediastinal structures.
- Structure: the right lung is divided into upper, middle and lower lobes by *oblique* and *horizontal fissures*. The left lung has only an oblique fissure, and hence no middle lobe. The *lingular segment* represents the left-sided equivalent of the right middle lobe. It is, however, an anatomical part of the left upper lobe.

Structures enter or leave the lungs by way of the lung hilum which, as mentioned earlier, is ensheathed in a loose pleural cuff (see Fig. 7.1).
- **Blood supply**: the bronchi and parenchymal tissue of the lungs are supplied by *bronchial arteries* – branches of the descending thoracic aorta. *Bronchial veins*, which also communicate with pulmonary veins, drain into the *azygos* and *hemiazygos*. The alveoli receive deoxygenated blood from terminal branches of the pulmonary artery and oxygenated blood returns via tributaries of the pulmonary veins. Two pulmonary veins return blood from each lung to the left atrium.
- **Lymphatic drainage of the lungs**: lymph returns from the periphery towards the hilar tracheobronchial groups of nodes and from here to mediastinal lymph trunks.
- **Nerve supply of the lungs**: a *pulmonary plexus* is located at the root of each lung. The plexus is composed of sympathetic fibres (from the sympathetic trunk – see p. 33) and parasympathetic fibres (from the vagus – see p. 33). Efferent fibres from the plexus supply the bronchial musculature and afferents are received from the mucous membranes of bronchioles and from the alveoli.

The mechanics of respiration

- A negative intrapleural pressure keeps the lungs continuously partially inflated.
- During normal *inspiration*: contraction of the upper external intercostals increases the antero-posterior (A-P) diameter of the upper thorax; contraction of the lower external intercostals increases the transverse diameter of the lower thorax; and contraction of the diaphragm increases the vertical length of the internal thorax. These changes serve to increase lung volume and thereby result in reduction of intrapulmonary pressure, causing air to be sucked into the lungs. In deep inspiration, the sternocleidomastoid, scalenus anterior and medius, serratus anterior and pectoralis major and minor all aid to maximise thoracic capacity. These muscles are, therefore, referred to collectively as the *accessory muscles of respiration*.
- *Expiration* is mostly due to passive relaxation of the muscles of inspiration and elastic recoil of the lungs. In forced expiration, the abdominal musculature aids ascent of the diaphragm.

The chest X-ray (CXR) (Fig. 9.6)

The standard CXR is the postero-anterior (P-A) view. This is taken with the subject's chest touching the cassette holder and the X-ray beam directed anteriorly from behind.

Structures visible on the CXR include:
- **Heart borders**: any significant enlargement of a particular chamber can be seen on the X-ray. In congestive cardiac failure, all four chambers of the heart are enlarged (*cardiomegaly*). This is identified on the P-A view as a cardiothoracic ratio greater than 0.5. This ratio is calculated by dividing the width of the heart by the width of the thoracic cavity at its widest point.
- **Lungs**: the lungs are radiolucent. Dense streaky shadows, seen at the lung roots, represent the blood-filled pulmonary vasculature.
- **Diaphragm**: the angle made between the diaphragm and chest wall is termed the *costophrenic angle*. This angle is lost when a pleural effusion collects.
- **Mediastinal structures**: these are difficult to distinguish as there is considerable overlap. Clearly visible, however, is the aortic arch which, when pathologically dilated (aneurysmal), creates the impression of 'widening' of the mediastinum.

9 The heart I

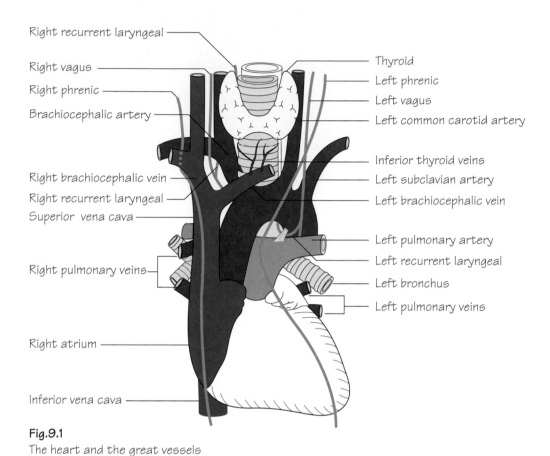

Right recurrent laryngeal

Right vagus

Right phrenic

Brachiocephalic artery

Right brachiocephalic vein

Right recurrent laryngeal

Superior vena cava

Right pulmonary veins

Right atrium

Inferior vena cava

Thyroid

Left phrenic

Left vagus

Left common carotid artery

Inferior thyroid veins

Left subclavian artery

Left brachiocephalic vein

Left pulmonary artery

Left recurrent laryngeal

Left bronchus

Left pulmonary veins

Fig.9.1
The heart and the great vessels

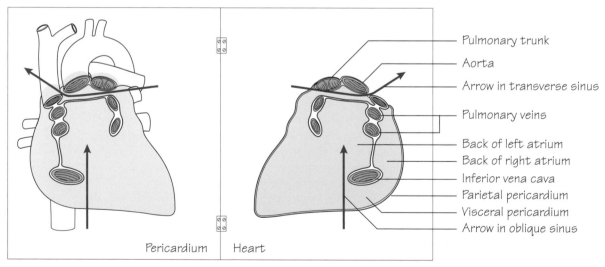

Pericardium   Heart

Pulmonary trunk

Aorta

Arrow in transverse sinus

Pulmonary veins

Back of left atrium

Back of right atrium

Inferior vena cava

Parietal pericardium

Visceral pericardium

Arrow in oblique sinus

Fig.9.2
The sinuses of the pericardium. The heart has been removed from the pericardial cavity and turned over to show its posterior aspect. The red line shows the cut edges where the visceral pericardium is continuous with the parietal pericardium. Visceral layer: blue, parietal layer: red

The middle mediastinum is comprised of the heart, pericardium, lung roots and the adjoining parts of the great vessels (Figs. 5.1 and 9.1).

The pericardium

The pericardium comprises fibrous and serous components. The *fibrous* pericardium is a strong layer that covers the heart. It fuses with the roots of the great vessels above and with the central tendon of the diaphragm below. The *serous* pericardium lines the fibrous pericardium (parietal layer) and is reflected at the vessel roots to cover the heart surface (visceral layer). The serous pericardium provides smooth surfaces for the heart to move against. Two important sinuses are located between the parietal and visceral layers. These are:

• **Transverse sinus**: located between pulmonary trunk and aorta anteriorly and the superior vena cava and left atrium posteriorly (Fig. 9.2).

• **Oblique sinus**: behind the left atrium. It is bounded by the inferior vena cava and the pulmonary veins (Fig. 9.2).

• *Blood supply*: from the pericardiacophrenic branches of the internal thoracic arteries.

• *Nerve supply*: the fibrous pericardium and the parietal layer of the serous pericardium are supplied by the phrenic nerve.

The heart surfaces

• The *anterior* (*sternocostal*) surface comprises the right atrium, atrioventricular groove, right ventricle, a small strip of left ventricle and the auricle of the left atrium.

• The *inferior* (*diaphragmatic*) surface comprises the right atrium, atrioventricular groove and both ventricles separated by the interventricular groove.

• The *posterior* surface (*base*) comprises the left atrium receiving the four pulmonary veins.

The heart chambers
The right atrium (Fig. 9.3)

• It receives deoxygenated blood from the inferior vena cava below and from the superior vena cava above.

• It receives the *coronary sinus* in its lower part (p. 31).

• The upper end of the atrium projects to the left of the superior vena cava as the *right auricle*.

• The *sulcus terminalis* is a vertical groove on the outer surface of the atrium. This groove corresponds internally to the *crista terminalis* – a muscular ridge which separates the smooth walled atrium (derived from the sinus venosus) from the rest of the atrium (derived from the true fetal atrium). The latter contains horizontal ridges of muscle – *musculi pectinati*.

• Above the coronary sinus the interatrial septum forms the posterior wall. The depression in the septum – the *fossa ovalis* – represents the site of the fetal *foramen ovale*. Its floor is the fetal *septum primum*. The upper ridge of the fossa ovalis is termed the *limbus*, which represents the *septum secundum*. Failure of fusion of the septum primum with the septum secundum gives rise to a patent foramen ovale (*atrial septal defect*) but, as long as the two septa still overlap, there will be no functional disability. A patent foramen may give rise to a left–right shunt (see Chapter 13).

The right ventricle

• It receives blood from the right atrium through the tricuspid valve (see below). The edges of the valve cusps are attached to *chordae tendineae* which are, in turn, attached below to papillary muscles. The latter are projections of muscle bundles on the ventricular wall.

• The wall of the right ventricle is thicker than that of both the atria, but not as thick as that of the left ventricle. The wall contains a mass of muscular bundles known as the *trabeculae carneae*. One prominent bundle projects forwards from the interventricular septum to the anterior wall. This is the *moderator band* (or septomarginal trabecula) and is of importance in the conduction of impulses as it contains the right branch of the atrioventricular bundle.

• The *infundibulum* is the smooth-walled outflow tract of the right ventricle.

• The pulmonary valve (see below) is situated at the top of the infundibulum. It is composed of three semilunar cusps. Blood flows through the valve and into the pulmonary arteries via the pulmonary trunk to be oxygenated in the lungs.

The left atrium

• It receives oxygenated blood from four pulmonary veins which drain posteriorly.

• The cavity is smooth walled except for the atrial appendage.

• On the septal surface a depression marks the fossa ovalis.

• The mitral (bicuspid) valve guards the passage of blood from the left atrium to the left ventricle.

The left ventricle (Fig. 9.4)

• The wall of the left ventricle is considerably thicker than that of the right ventricle, but the structure is similar. The thick wall is necessary to pump oxygenated blood at high pressure through the systemic circulation. Trabeculae carneae project from the wall with papillary muscles attached to the mitral valve cusp edges by way of chordae tendineae.

• The *vestibule* is a smooth-walled part of the left ventricle which is located below the aortic valve and constitutes the outflow tract.

The heart valves (Fig. 9.5)

• The purpose of valves within the heart is to maintain unidirectional flow.

• The *mitral* (*bicuspid*) and *tricuspid* valves are flat. During ventricular systole, the free edges of the cusps come into contact and eversion is prevented by the pull of the chordae.

• The *aortic* and *pulmonary* valves are composed of three semilunar cusps which are cup-shaped. During ventricular diastole, back-pressure of blood above the cusps forces them to fill and hence close.

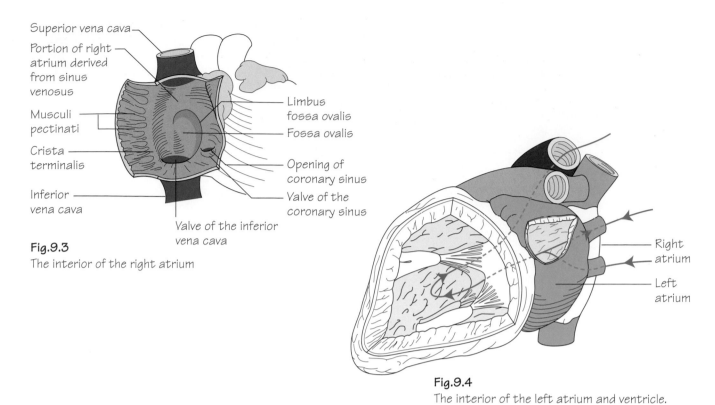

Fig.9.3
The interior of the right atrium

Superior vena cava

Portion of right atrium derived from sinus venosus

Musculi pectinati

Crista terminalis

Inferior vena cava

Valve of the inferior vena cava

Limbus fossa ovalis

Fossa ovalis

Opening of coronary sinus

Valve of the coronary sinus

Fig.9.4
The interior of the left atrium and ventricle. The arrow shows the direction of blood flow. Note that blood flows over both surfaces of the anterior cusp of the mitral valve

Right atrium

Left atrium

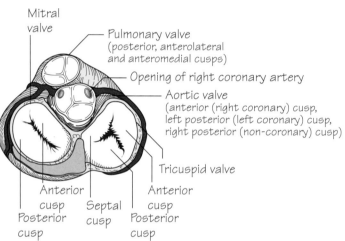

Mitral valve

Pulmonary valve (posterior, anterolateral and anteromedial cusps)

Opening of right coronary artery

Aortic valve (anterior (right coronary) cusp, left posterior (left coronary) cusp, right posterior (non-coronary) cusp)

Tricuspid valve

Anterior cusp

Septal cusp

Anterior cusp

Posterior cusp

Posterior cusp

Fig.9.5
A section through the heart at the level of the valves. The aortic and pulmonary valves are closed and the mitral and tricuspid valves open, as they would be during ventricular diastole

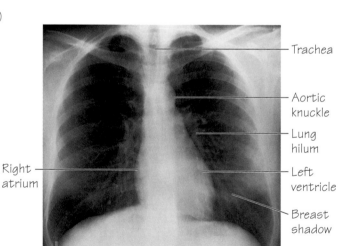

Trachea

Aortic knuckle

Lung hilum

Left ventricle

Breast shadow

Right atrium

Costophrenic angle

Fig.9.6
Normal postero-anterior chest X-ray (see p. 25)

Clinical notes

- **Cardiac tamponade**: following thoracic trauma, blood can collect in the pericardial space (haemopericardium) which may, in turn, lead to cardiac tamponade. This manifests itself clinically as shock, distended neck veins and muffled/absent heart sounds (Beck's triad). This condition is fatal unless pericardial decompression is effected immediately.
- **Valvular disease of the heart and cardiac murmurs**: numerous pathological processes affect the heart valves to cause either thickening of the cusps with resultant stenosis or rupture of the valvular mechanism with consequent regurgitation.
  - *Mitral stenosis*: is commonly associated with a previous history of rheumatic fever. On auscultation, a loud opening snap can often be heard in early diastole. This represents the opening of the mitral valve. In addition, a mid-diastolic murmur is frequently present. The latter occurs as a result of turbulent flow across the stenotic valve during ventricular filling.
  - *Mitral regurgitation*: numerous disease processes can result in compromised mitral valve integrity. An acute cause arises due to rupture of the chordae tendineae following myocardial infarction. Mitral regurgitation is evident clinically by auscultation of a pansystolic murmur. The latter represents regurgitant flow of blood from the ventricle to the atrium during systole.
  - *Aortic stenosis*: occurs mostly as a result of arteriosclerotic degeneration of the valve or congenital valvular abnormality. Classically, this condition is characterised by a low-volume pulse in association with an ejection systolic murmur.
  - *Aortic regurgitation*: numerous conditions give rise to aortic regurgitation. Clinical manifestations of this valvular dysfunction include an increase in the pulse pressure (*water-hammer pulse*) in association with a high-pitched early diastolic murmur.

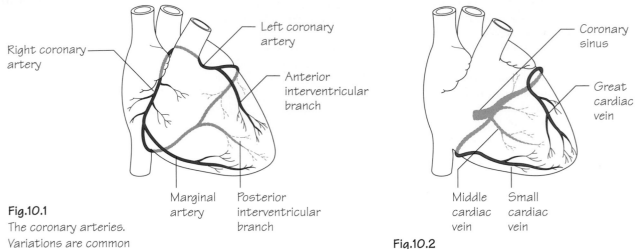

Fig.10.1
The coronary arteries.
Variations are common

Right coronary artery

Left coronary artery

Anterior interventricular branch

Marginal artery

Posterior interventricular branch

Coronary sinus

Great cardiac vein

Middle cardiac vein

Small cardiac vein

Fig.10.2
The venous drainage of the heart

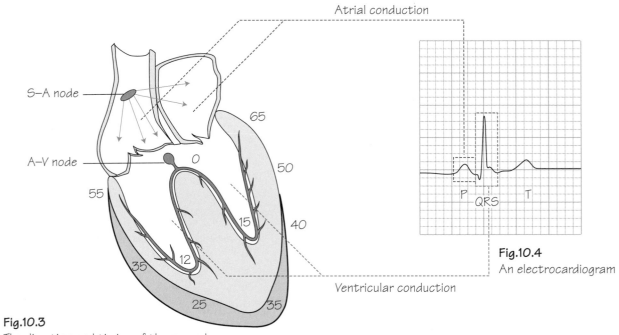

Atrial conduction

S–A node

A–V node

65

50

55

0

15

40

35

12

25

35

Ventricular conduction

P QRS T

Fig.10.4
An electrocardiogram

Fig.10.3
The direction and timing of the spread
of action potential in the conducting
system of the heart.
Times are in msec

30 *Anatomy at a Glance*, Third Edition. Omar Faiz, Simon Blackburn and David Moffat.
© 2011 Blackwell Publishing Ltd. Published 2011 by Blackwell Publishing Ltd.

The arterial supply of the heart
(Fig. 10.1)

The coronary arteries are responsible for supplying the heart itself with oxygenated blood.

The origins of the coronary arteries are as follows:
- The left coronary artery arises from the aortic sinus immediately above the left posterior cusp of the aortic valve (see Fig. 9.5).
- The right coronary artery arises from the aortic sinus immediately above the anterior cusp of the aortic valve (see Fig. 9.5).

The course of the coronary arteries and their principal branches is illustrated diagrammatically in Fig. 10.1. For the most part, the principal coronary vessels traverse the heart between the major chambers (i.e. within the atrioventricular groove and interventricular sulcus). The latter probably represent the sites of least stretch and, in consequence, least impedance to flow.

There is considerable variation in the size and distribution zones of the coronary arteries. For example, in some people, the *posterior interventricular* branch of the right coronary artery is large and supplies a large part of the left ventricle, whereas in the majority this is supplied by the *anterior interventricular* branch of the left coronary artery.

Similarly, the *sinu-atrial node* is usually supplied by a nodal branch of the right coronary artery but, in 30–40% of the population, it receives its supply from the left coronary artery. The A-V node is supplied by the right coronary artery in 90% of subjects, and the left coronary artery in the remaining 10%.

The venous drainage of the heart
(Fig. 10.2)

The venous drainage systems in the heart include:
- The *veins which accompany the coronary arteries* and drain into the right atrium via the *coronary sinus*. The coronary sinus drains into the right atrium to the left of and superior to the opening of the inferior vena cava. The *great cardiac vein* follows the anterior interventricular branch of the left coronary artery and then sweeps backwards to the left in the atrioventricular groove. The *middle cardiac vein* follows the posterior interventricular artery and, along with the *small cardiac vein* which follows the marginal artery, drains into the coronary sinus. The coronary sinus drains the vast majority of the heart's venous blood.
- The *venae cordis minimi*: these are small veins which drain directly into the cardiac chambers.
- The *anterior cardiac veins*: these are small veins which cross the atrioventricular groove to drain directly into the right atrium.

The conducting system of the heart
(Figs. 10.3 and 10.4)

- The sinu-atrial (S-A) node is the pacemaker of the heart. It is situated near the top of the crista terminalis, below the superior vena caval opening into the right atrium. Impulses generated by the S-A node are conducted throughout the atrial musculature to effect synchronous atrial contraction. Disease or degeneration of any part of the conduction pathway can lead to dangerous interruption of heart rhythm. Degeneration of the S-A node leads to other sites of the conduction pathway taking over the pacemaking role, albeit usually at a slower rate.
- Impulses reach the *atrioventricular* (A-V) node which lies in the interatrial septum, just above the opening for the coronary sinus. From here, the impulse is transmitted to the ventricles via the *atrioventricular bundle (of His)*, which descends in the interventricular septum.
- The *bundle of His* divides into right and left branches which send *Purkinje fibres* to lie within the subendocardium of the ventricles. The position of the Purkinje fibres accounts for the almost synchronous contraction of the ventricles.

The nerve supply of the heart

The heart receives both a sympathetic and a parasympathetic nerve supply so that the heart rate can be controlled to demand.
- The *parasympathetic supply* (bradycardic effect) is derived from the vagus nerve (p. 33).
- The *sympathetic supply* (tachycardic and positively inotropic effect) is derived from the cervical and upper thoracic sympathetic ganglia by way of superficial and deep cardiac plexuses (p. 33).

Clinical notes

- **Ischaemic heart disease**: the coronary arteries are functional end-arteries. Following a total occlusion, therefore, the myocardium supplied by the blocked artery is deprived of its blood supply (myocardial infarction). When the vessel lumen gradually narrows due to atheromatous change of the walls, patients complain of gradually increasing chest pain on exertion (angina). Under these conditions, the increased demand placed on the myocardium cannot be met by the diminished arterial supply. Angina that is not amenable to pharmacological control can be relieved by dilating (angioplasty), or surgically bypassing (coronary artery bypass grafting) the arterial stenosis. The latter procedure is usually performed using a reversed length of great saphenous vein anastomosed to the proximal aorta and then distally to the coronary artery beyond the stenosis. The internal thoracic artery may also be used with its distal end divided and anastomosed to a coronary artery distal to the stenosis. Ischaemic heart disease is the leading cause of death in the Western world and, consequently, a thorough knowledge of the coronary anatomy is essential.

11 The nerves of the thorax

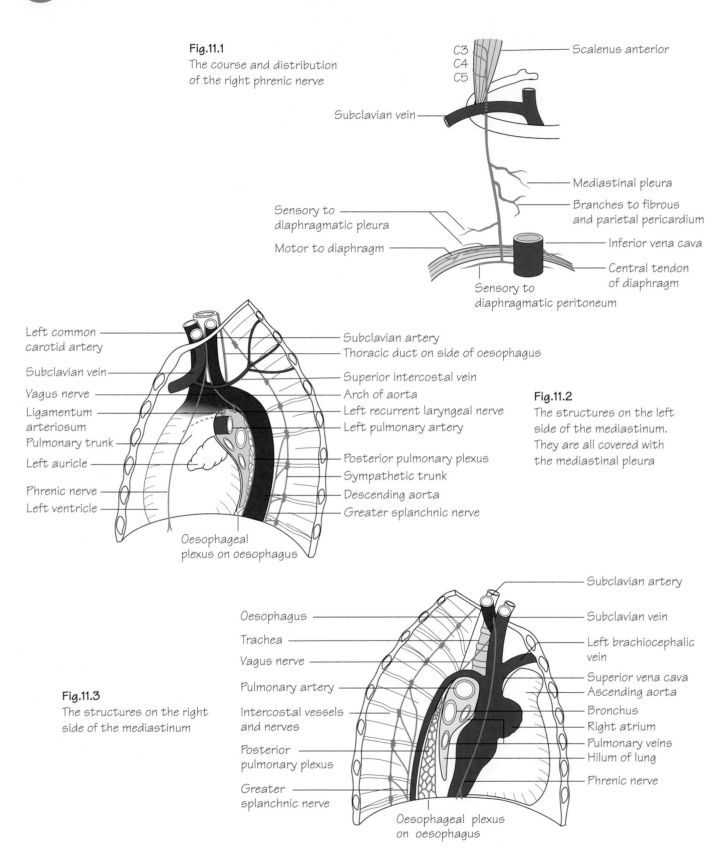

Fig.11.1

The course and distribution of the right phrenic nerve

C3
C4
C5

Scalenus anterior

Subclavian vein

Mediastinal pleura

Branches to fibrous and parietal pericardium

Sensory to diaphragmatic pleura

Motor to diaphragm

Inferior vena cava

Central tendon of diaphragm

Sensory to diaphragmatic peritoneum

Left common carotid artery

Subclavian vein

Vagus nerve

Ligamentum arteriosum

Pulmonary trunk

Left auricle

Phrenic nerve

Left ventricle

Subclavian artery

Thoracic duct on side of oesophagus

Superior intercostal vein

Arch of aorta

Left recurrent laryngeal nerve

Left pulmonary artery

Posterior pulmonary plexus

Sympathetic trunk

Descending aorta

Greater splanchnic nerve

Oesophageal plexus on oesophagus

Fig.11.2

The structures on the left side of the mediastinum. They are all covered with the mediastinal pleura

Fig.11.3

The structures on the right side of the mediastinum

Oesophagus

Trachea

Vagus nerve

Pulmonary artery

Intercostal vessels and nerves

Posterior pulmonary plexus

Greater splanchnic nerve

Oesophageal plexus on oesophagus

Subclavian artery

Subclavian vein

Left brachiocephalic vein

Superior vena cava

Ascending aorta

Bronchus

Right atrium

Pulmonary veins

Hilum of lung

Phrenic nerve

The phrenic nerves

The phrenic nerves arise from the C3, C4 and C5 nerve roots in the neck.

- The *right phrenic nerve* (Fig. 11.1) descends along a near-vertical path, anterior to the lung root, lying on the right brachiocephalic vein, the superior vena cava and the right atrium sequentially, before passing to the inferior vena caval opening in the diaphragm (T8). Here, the right phrenic enters the caval opening and immediately penetrates the diaphragm which it innervates from its inferior surface.

- The *left phrenic nerve* (Fig. 11.2) descends alongside the left subclavian artery. On the arch of the aorta it passes over the left superior intercostal vein to descend in front of the left lung root onto the pericardium overlying the left ventricle. The left phrenic nerve then pierces the muscular diaphragm as a solitary structure before being distributed on its inferior surface. It should be noted that the phrenic nerves do not pass beyond the undersurface of the diaphragm.

- The phrenic nerves are composed mostly of motor fibres which supply the diaphragm. However, they also transmit fibres that are sensory to the fibrous pericardium, mediastinal pleura and peritoneum as well as the central part of the diaphragm. Irritation of the diaphragmatic peritoneum is usually referred to the C4 dermatome. Upper abdominal pathology, such as a perforated duodenal ulcer, therefore, often results in shoulder tip pain.

The vagi

The vagi are the 10th cranial nerves (p. 145).

- The *right vagus nerve* (Figs. 5.2 and 11.3) descends adherent to the thoracic trachea, prior to passing behind the lung root to form the posterior pulmonary plexus. It finally reaches the lower oesophagus, where it forms an oesophageal plexus with the left vagus. From this plexus, anterior and posterior vagal trunks descend (carrying fibres from both left and right vagi) on the oesophagus to pass into the abdomen through the oesophageal opening in the diaphragm at the level of T10.

- The left vagus nerve (Fig. 11.2) crosses the arch of the aorta and its branches. It is itself crossed here by the left superior intercostal vein. Below, it descends behind the lung root to reach the oesophagus where it contributes to the oesophageal plexus mentioned above (see Fig. 5.2).

Vagal branches

- The *left recurrent laryngeal nerve* arises from the left vagus below the arch of the aorta. It hooks around the ligamentum arteriosum and ascends in the groove between the trachea and the oesophagus to reach the larynx (p. xx).

- The *right recurrent laryngeal nerve* arises from the right vagus in the neck and hooks around the right subclavian artery prior to ascending in the groove between the trachea and the oesophagus, before finally reaching the larynx.

- The recurrent laryngeal nerves supply the mucosa of the upper trachea and oesophagus, as well as providing a motor supply to all of the muscles of the larynx (except cricothyroid) and sensory fibres to the lower larynx.

- The vagi also contribute branches to the cardiac and pulmonary plexuses.

The thoracic sympathetic trunk (Figs. 11.2 and 11.3, and Chapter 58)

- The *thoracic sympathetic chain* is a continuation of the cervical chain. It descends in the thorax behind the pleura immediately lateral to the vertebral bodies and passes under the medial arcuate ligament of the diaphragm to continue as the *lumbar sympathetic trunk*.

- The thoracic chain bears a ganglion for each spinal nerve; the first frequently joins the inferior cervical ganglion to form the *stellate ganglion*. Each ganglion receives a white ramus communicans, containing preganglionic fibres from its corresponding spinal nerve, and sends back a grey ramus bearing postganglionic fibres.

Branches

- Sympathetic fibres are distributed to the skin with each of the thoracic spinal nerves.

- Postganglionic fibres from T1–5 are distributed to the thoracic viscera – the heart and the great vessels, the lungs and the oesophagus.

- Preganglionic fibres from T5–12 form the *splanchnic nerves*. These pierce the crura of the diaphragm and pass to the coeliac and renal ganglia, from which they are relayed as postganglionic fibres to the abdominal viscera (cf. fibres to the suprarenal medulla which are preganglionic). These splanchnic nerves are the *greater splanchnic* (T5–10), *lesser splanchnic* (T10–11) and *least splanchnic* (T12). They lie medial to the sympathetic trunk on the bodies of the thoracic vertebrae and are quite easily visible through the parietal pleura.

The cardiac plexus

This plexus is, for descriptive purposes, divided into superficial and deep parts. It consists of sympathetic and parasympathetic efferents as well as afferents.

- Cardiac branches from the plexus innervate the heart; they accompany the coronary arteries for vasomotor control and supply the sinuatrial and atrioventricular nodes, via which they control the heart rate.

- Pulmonary branches innervate the bronchial wall smooth muscle, controlling diameter, and pulmonary blood vessels, for vasomotor control.

Clinical notes

- *Sympathectomy*: upper limb sympathectomy is used for the treatment of hyperhidrosis and Raynaud's syndrome. Surgical sympathectomy involves excision of part of the thoracic sympathetic chain below the level of the stellate ganglion. The latter structure must be identified on the neck of the 1st rib and preserved. Injury to the stellate ganglion may result in an ipsilateral Horner's syndrome.

12 Surface anatomy of the thorax

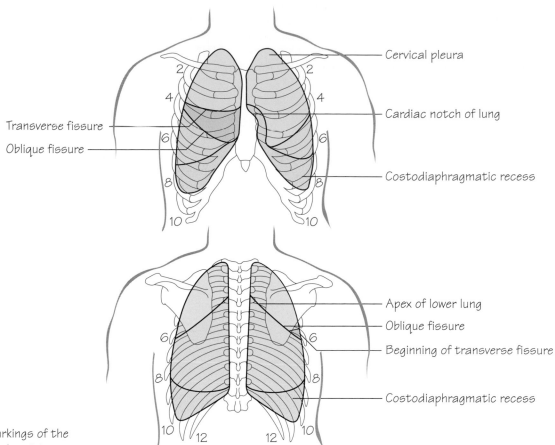

Cervical pleura

Cardiac notch of lung

Costodiaphragmatic recess

Transverse fissure

Oblique fissure

Apex of lower lung

Oblique fissure

Beginning of transverse fissure

Costodiaphragmatic recess

Fig.12.1
The surface markings of the
lungs and pleural cavities

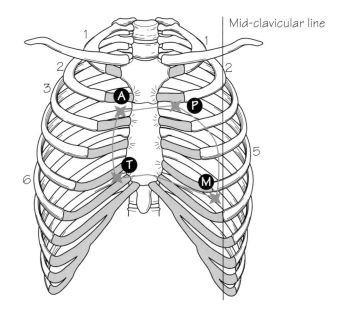

Mid-clavicular line

Fig.12.2
The surface markings of the heart.
The areas of auscultation for the
aortic, pulmonary, mitral and
tricuspid valves are indicated by letters

Anatomy at a Glance, Third Edition. Omar Faiz, Simon Blackburn and David Moffat.
© 2011 Blackwell Publishing Ltd. Published 2011 by Blackwell Publishing Ltd.

The anterior thorax

Landmarks of the anterior thorax include:

• The *manubriosternal angle (angle of Louis)*: formed by the joint between the manubrium and body of the sternum. It is an important landmark as the 2nd costal cartilages articulate on either side and, by following this line onto the 2nd rib, further ribs and intercostal spaces can be identified. The sternal angle corresponds to a horizontal point level with the intervertebral disc between T4 and T5.

• The *suprasternal notch*: situated in the midline between the medial ends of the clavicles and above the upper edge of the manubrium.

• The *costal margin*: formed by the lower borders of the cartilages of the 7th, 8th, 9th and 10th ribs and the ends of the 11th and 12th ribs.

• The *xiphisternal joint*: formed by the joint between the body of the sternum and xiphisternum.

The posterior thorax

Landmarks of the posterior thorax include:

• The first palpable spinous process of C7 (*vertebra prominens*). The C1–6 vertebrae are covered by the thick ligamentum nuchae. The spinous processes of the thoracic vertebrae can be palpated and counted in the midline posteriorly.

• The scapula is located on the upper posterior chest wall. In slim subjects, the superior angle, inferior angle, spine and medial (vertebral) border of the scapula are easily palpable.

Lines of orientation

These are imaginary vertical lines used to describe locations on the chest wall. They include:

• The *mid-clavicular line*: a vertical line from the midpoint of the clavicle downwards.

• The *anterior* and *posterior axillary lines*: from the anterior and posterior axillary folds, respectively, vertically downwards.

• The *mid-axillary line*: from the midpoint between anterior and posterior axillary lines vertically downwards.

The surface markings of thoracic structures

The trachea

The trachea commences at the lower border of the cricoid cartilage (C6 vertebral level). It runs downwards in the midline and ends slightly to the right by bifurcating into the left and right main bronchi. The bifurcation occurs at the level of the sternal angle (T4/5).

The pleura (Fig. 12.1)

The apex of the pleura projects about 2.5 cm above the medial third of the clavicle. The lines of pleural reflection pass behind the sternoclavicular joints to meet in the midline at the level of the sternal angle. The right pleura then passes downwards to the 6th costal cartilage. The left pleura passes laterally for a small distance at the 4th costal cartilage and descends vertically lateral to the sternal border to the 6th costal cartilage. From these points, both pleurae pass posteriorly and, in doing so, cross the 8th rib in the mid-clavicular line, the 10th rib in the mid-axillary line and, finally, reach the level of the 12th rib posteriorly.

The lungs (Fig. 12.1)

The apex and mediastinal border of the right lung follow the pleural outline. In mid-inspiration, the right lung lower border crosses the 6th rib in the mid-clavicular line, the 8th rib in the mid-axillary line and reaches the level of the 10th rib posteriorly. The left lung borders are similar to those of the right except that the mediastinal border arches laterally (the cardiac notch), but then resumes the course mentioned above.

• The *oblique fissure*: is represented by an oblique line drawn from a point 2.5 cm lateral to the 5th thoracic spinous process to the 6th costal cartilage anteriorly. The oblique fissures separate the lungs into upper and lower lobes.

• The *transverse fissure*: is represented by a line drawn horizontally from the 4th costal cartilage to a point where it intersects the oblique fissure. The fissure separates the upper and middle lobes of the right lung.

The heart

• The borders of the heart are illustrated by joining the four points as shown (Fig. 12.2).

• The apex of the left ventricle corresponds to where the apex beat is palpable. The surface marking for the apex beat is in the 5th intercostal space in the mid-clavicular line.

• See Fig. 12.2 for optimal sites of valvular auscultation.

The great vessels

• The *aortic arch*: arches antero-posteriorly behind the manubrium. The highest point of the arch reaches the midpoint of the manubrium.

• The *brachiocephalic artery* and *left common carotid artery*: ascend posterior to the manubrium.

• The *brachiocephalic veins*: formed by the confluence of the internal jugular and subclavian veins which occurs posterior to the sternoclavicular joints.

• The *superior vena cava*: formed by the confluence of the left and right brachiocephalic veins between the 2nd and 3rd right costal cartilages at the right border of the sternum.

The breast

The base of the breast (p. 83) is constant, overlying the 2nd to the 6th ribs and costal cartilages anteriorly and from the lateral border of the sternum to the mid-axillary line. The position of the nipple is variable in the female, but in the male is usually in the 4th intercostal space in the mid-clavicular line.

The internal thoracic vessels

These vessels (arteries and veins) descend 1 cm lateral to the edge of the sternum.

The diaphragm

In mid-inspiration, the highest part of the right dome reaches as far as the upper border of the 5th rib in the mid-clavicular line. The left dome reaches only the lower border of the 5th rib.

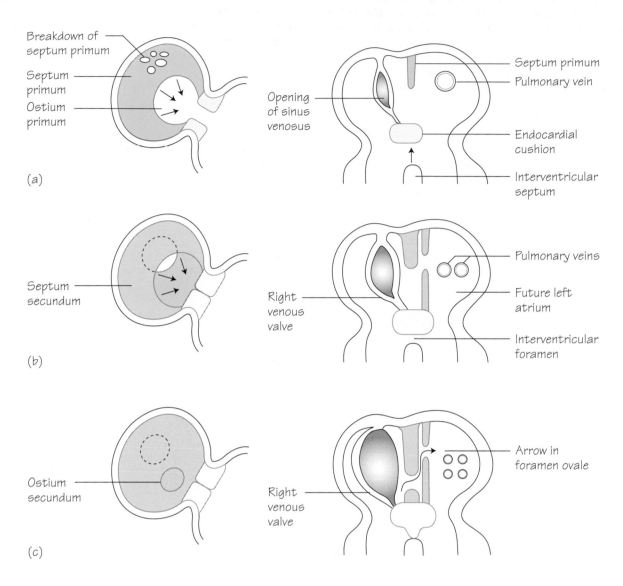

Breakdown of septum primum

Septum primum

Ostium primum

(a)

Opening of sinus venosus

Septum primum

Pulmonary vein

Endocardial cushion

Interventricular septum

Septum secundum

(b)

Right venous valve

Pulmonary veins

Future left atrium

Interventricular foramen

Ostium secundum

(c)

Right venous valve

Arrow in foramen ovale

Fig.13.1
Three stages in the development of the interatrial septum. The left-hand diagrams show the right side of the interatrial septum and the right-hand diagrams show a coronal section through the two atria

Development of the heart

The embryonic heart tube consists of the *sinus venosus*, the *atrium*, the *ventricle* and the *truncus arteriosus*. The opening of the sinus venosus into the atrium is guarded by right and left *venous valves* which project into the atrium. A pair of endocardial *atrioventricular cushions* partly occludes the passage between the atrium and ventricle.

• *Septation of the atria*. The crescentic *septum primum* grows down, leaving the *ostium primum* below its free border (Fig. 13.1). Before it reaches the atrioventricular cushions, however, a hole appears in its upper part – the *ostium secundum*. The thicker *septum secundum* grows down on the right side of the septum primum and overlaps the ostium secundum so that the opening between them, the *foramen ovale*, takes the form of a flap-valve, allowing blood to pass from right to left but not from left to right. The two septa fuse after birth. The sinus venosus is taken into the right atrium to form the smooth part of the atrium behind the crista terminalis which, itself, is formed from the right venous valve (p. 27).

• *Septation of the ventricle*. A thick *interventricular septum* grows up towards the atrioventricular cushions, leaving an *interventricular foramen* above its free border. The endocardial cushions fuse centrally so that there are separate right and left atrioventricular openings, and the interventricular foramen is closed by growth of tissue from the cushions.

• The pulmonary veins from the lungs open into the left atrium, at first by a single opening, but the veins are then taken into the atrium together with their first branchings, so that eventually there are four openings.

• The truncus arteriosus is divided into the aorta and pulmonary trunks by the development of a *spiral septum*, so that the aorta communicates with the left ventricle and the pulmonary trunk with the right.

Developmental anomalies of the heart

Although there are a huge number of well-recognised anomalies of heart development, only the most common are mentioned here.

• **Ventricular septal defect**. This is the most common defect and results from failure of closure of the interventricular foramen. If the defect is small, it may be symptomless, and the foramen may close spontaneously. Larger defects will produce a large left–right shunt after birth when the pulmonary pressure drops, decreasing the right heart pressure. This may lead to heart failure and surgical treatment may be required.

• **Ostium primum defect**. There is failure of the septum primum to reach the endocardial cushions which, themselves, fail to develop properly. There are, thus, both atrial and ventricular septal defects, together with maldevelopment of the mitral and tricuspid valves. About one-third of babies with this defect also suffer from Down's syndrome. Surgical repair is difficult.

• **Ostium secundum defect**. In its mildest form, there is failure of fusion between the two septa, but they still overlap (*probe patency*). The septum primum is kept pressed against the septum secundum, so that the condition is symptomless unless, for some reason, the pressure rises on the right side of the heart in later life. If there is no overlap, there will be a left–right shunt but, again, this may be symptomless.

• **Tetralogy of Fallot**. This is the result of four simultaneous defects: a *ventricular septal defect*, *an aorta which overrides the free upper border of the interventricular septum*, *pulmonary stenosis* (narrowing in the region of the pulmonary valve) and *right ventricular hypertrophy*. The pulmonary stenosis causes a rise in pressure in the right ventricle, so that there is a right–left shunt through the ventricular septal defect, resulting in cyanosis. Surgical treatment is possible and the prognosis is excellent.

• **Transposition of the great vessels**. Because of the defective development of the spiral septum, the aorta and pulmonary trunks are reversed in position, so that the aorta communicates with the right ventricle and the pulmonary trunk with the left ventricle. The condition leads to death if left untreated. Shunting of oxygenated blood from right to left can be produced by making a deliberate perforation in the interatrial septum, thereby permitting deferred definitive surgical correction.

It must be stressed that the above is an anatomical classification of cardiac anomalies. Clinically, a different classification is used depending on the age of onset of symptoms and the presence or absence of cyanosis and/or heart failure.

Development of the air passages

• The trachea begins development as a *laryngo-tracheal groove* in the floor of the primitive pharynx. It separates itself from the developing oesophagus, except at its upper end, and grows distally, dividing into two lung buds which invaginate the pleural cavities. Further subdivision takes place to form the bronchi and smaller air passages and, ultimately, the alveoli. The cartilages of the larynx develop from the 4th and 6th branchial arches around the upper end of the trachea.

• Alveolar development commences during intra-uterine life and continues after birth. During the last 3 months of intra-uterine life, the smallest subdivisions of the respiratory tree can be recognized as respiratory bronchioles with primitive alveoli opening off them. In the last few weeks before birth, the alveolar Type 2 cells secrete a phospholipid material, known as *surfactant*, whose function is to lower the surface tension of the fluid in the alveoli so that they do not collapse during expiration. Some breathing movements occur before birth, but the lungs do not function because they are filled with fluid and, because of the open ductus arteriosus, the pulmonary blood flow is very low. At birth, the pulmonary circulation opens up, the fluid in the lungs is rapidly absorbed and the alveoli dilate and become air filled. The formation of new alveoli continues as the lungs enlarge, up to the age of about 8 years.

Developmental anomalies of the air passages

• **Tracheo-oesophageal fistula**. This is the most common anomaly of the upper respiratory tract and is usually associated with *oesophageal atresia* (complete obstruction or absence of a segment of the oesophagus). In the most common variety of this anomaly, the oesophagus ends blindly as it enters the thorax, and there is a communication between the lower segment of the oesophagus and the trachea just above the carina. If the baby is fed before the diagnosis is made, the dilated upper segment of the oesophagus may spill over into the air passages. In addition, positive pressure ventilation causes the stomach to become dilated via the fistula, causing respiratory compromise. Surgical treatment is necessary to correct the condition.

• **Respiratory distress syndrome (hyaline membrane disease)**. This is caused by deficiency of surfactant in premature babies. The alveoli do not expand properly owing to the surface tension of the fluid in the alveoli, so that full oxygenation of the blood cannot occur and the baby suffers respiratory distress. The condition requires the administration of exogenous surfactant derived from animal tissues as well as ventilatory support in many cases.

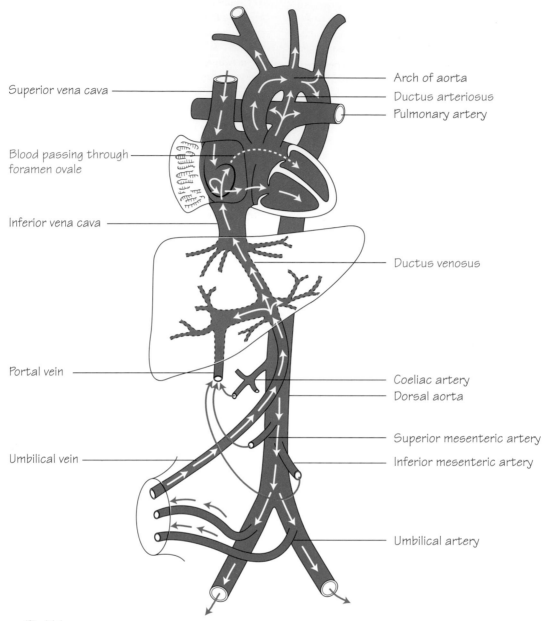

Superior vena cava

Blood passing through
foramen ovale

Inferior vena cava

Portal vein

Umbilical vein

Arch of aorta
Ductus arteriosus
Pulmonary artery

Ductus venosus

Coeliac artery
Dorsal aorta

Superior mesenteric artery
Inferior mesenteric artery

Umbilical artery

Fig.14.1
The fetal circulation. The three 'short circuits' are labelled in red

Because the fetal lungs are functionless, the essentials for the fetal circulation (Fig. 14.1) are that oxygenated blood and nutrients are passed from the placenta via the umbilical vein to the left side of the heart as directly as possible and, to this end, there are three short circuits or shunts:

1 The *ductus venosus* which bypasses the liver and directs most of the blood from the umbilical vein into the inferior vena cava.

2 The *foramen ovale* which shunts most of the blood coming up the inferior vena cava into the left atrium and thence to the left ventricle.

3 The *ductus arteriosus* which passes from the left pulmonary artery near its origin from the pulmonary trunk to the aorta. It, thus, bypasses the lungs and directs most of the blood in the pulmonary arteries to the aorta.

The blood in the umbilical vein is oxygenated in the placenta. It passes in the umbilical vein from the umbilicus to the region of the future porta hepatis where it joins the developing left branch of the portal vein. Thus, some of this blood passes into the hepatic sinusoids, but most of it is diverted through the wide *ductus venosus* into the inferior vena cava, where it mixes with venous blood.

Upon entering the right atrium, the stream of blood encounters the free border of the septum secundum which directs most of the blood through the *foramen ovale* into the left atrium and thence to the left ventricle. The rest of the blood remains in the right atrium and is mixed with the venous blood returning to the heart from the head and upper limbs via the superior vena cava; it then passes into the right ventricle and thence to the pulmonary trunk.

The lungs are in a collapsed condition and the pulmonary vessels are constricted and tightly coiled, so that there is a high vascular resistance. Most of the blood in the pulmonary arteries is, therefore, diverted straight into the aorta via the *ductus arteriosus*, which is a large vessel almost as big as the aorta. Because of this, the wall of the fetal right ventricle is as thick as that of the left, or even thicker; it is not until after birth that the wall of the left ventricle becomes the thicker of the two.

Blood in the descending aorta is, thus, in a poorer state of oxygenation than that in the aortic arch, so that the brain and upper limbs receive the best oxygenated blood, which is important for the rapid development of the brain. Some of the blood in the descending aorta is distributed to the abdomen and lower limbs, but the majority is passed into the two umbilical arteries and thence back to the placenta.

Changes in the fetal circulation at birth

Immediately after birth, the thick muscular walls of the umbilical vessels contract and the pulsation in the umbilical arteries can no longer be felt. The cord can then be tied off. The ductus venosus also closes down; this process possibly starts even before birth and continues for some time after. With the baby's first breaths, the lungs expand, the pulmonary vessels uncoil and most of the output of the right ventricle passes into the lungs. The ductus arteriosus also closes off as a result of contraction of the thick layer of smooth muscle in its walls and of the decreased resistance to the pulmonary blood flow. The consequent increase in venous return from the lungs causes pressure to rise in the left atrium, and the thin and flexible septum primum is pressed against the septum secundum, thus, closing off the foramen ovale.

These changes are, at first, physiological rather than anatomical, and it is only later that the changes become permanent.

The umbilical vein is finally obliterated and becomes a fibrous band, the *ligamentum teres*, passing from the umbilicus to the left branch of the portal vein and lying in the free border of the falciform ligament. Similarly, the ductus venosus becomes the fibrous *ligamentum venosum* that joins the left branch of the portal vein to the inferior vena cava. The proximal parts of the two umbilical arteries remain patent and form the stems of the obturator arteries, but the remainder is converted into the fibrous *medial umbilical ligaments* that ascend from the pelvis to the umbilicus.

• The ductus arteriosus closes off gradually, but some blood may continue to flow in it for up to 2 weeks after birth. It then becomes the fibrous *ligamentum arteriosum* that joins the left pulmonary artery to the underside of the arch of the aorta. As the ductus is developed from the left 6th aortic arch of the embryo, the left recurrent laryngeal nerve hooks around it before ascending to the larynx.

Developmental anomalies

• **Patent foramen ovale**. As mentioned above, anatomical closure of the foramen ovale is not immediate and, even when normally developed, the foramen maintains its *probe patency* for some time, possibly for a lifetime. The baby's crying causes a rise in venous pressure, so that blood may be shunted from the right atrium to the left, giving rise to temporary cyanosis; however, as long as the two septa overlap, there is no permanent effect.

• **Patent ductus arteriosus**. The ductus arteriosus is kept open by prostaglandin E_1 (PGE_1). In the presence of hypoxia, for example in premature babies with respiratory distress syndrome, PGE_1 is produced locally and prevents the ductus arteriosus from closing. The high pulmonary vascular resistance causes blood to shunt through the ductus arteriosus into the aorta, which further increases hypoxia. *Indomethacin*, which interferes with the production of prostaglandin synthetase, may cause the duct to close, but ultimately surgical closure may be necessary.

• **Coarctation of the aorta**. This is a congenital narrowing of the aorta in the region of the ductus arteriosus. It is possibly caused by the extension of the process of closure of the ductus into the aorta itself, although this does not explain its occasional occurrence more distally, such as near the origin of the renal arteries. When severe, it presents in infancy, often when the ductus arteriosus closes, and surgical correction is urgent. In the less severe type, the blood is passed to the lower part of the body by a collateral circulation involving the branches of the subclavian artery and the vessels of the chest wall. It may be symptomless and is often detected at a school medical examination. Surgical treatment is possible.

15 The abdominal wall

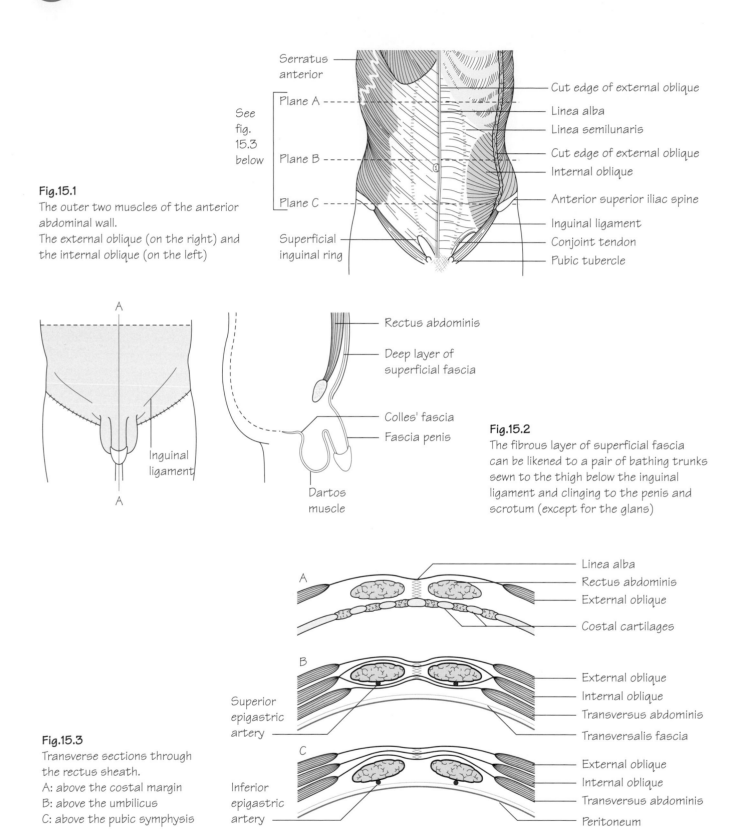

Fig.15.1
The outer two muscles of the anterior abdominal wall.
The external oblique (on the right) and the internal oblique (on the left)

Serratus anterior
Cut edge of external oblique
Linea alba
Linea semilunaris
Cut edge of external oblique
Internal oblique
Anterior superior iliac spine
Inguinal ligament
Conjoint tendon
Pubic tubercle
Superficial inguinal ring
See fig. 15.3 below
Plane A
Plane B
Plane C

Inguinal ligament

Rectus abdominis
Deep layer of superficial fascia
Colles' fascia
Fascia penis
Dartos muscle

Fig.15.2
The fibrous layer of superficial fascia can be likened to a pair of bathing trunks sewn to the thigh below the inguinal ligament and clinging to the penis and scrotum (except for the glans)

Fig.15.3
Transverse sections through the rectus sheath.
A: above the costal margin
B: above the umbilicus
C: above the pubic symphysis

Superior epigastric artery
Inferior epigastric artery

Linea alba
Rectus abdominis
External oblique
Costal cartilages

External oblique
Internal oblique
Transversus abdominis
Transversalis fascia

External oblique
Internal oblique
Transversus abdominis
Peritoneum

 Anatomy at a Glance, Third Edition. Omar Faiz, Simon Blackburn and David Moffat.

(a)

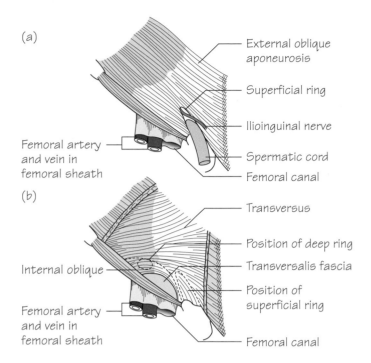

External oblique aponeurosis

Superficial ring

Ilioinguinal nerve

Spermatic cord

Femoral canal

Femoral artery and vein in femoral sheath

(b)

Transversus

Position of deep ring

Transversalis fascia

Position of superficial ring

Femoral canal

Internal oblique

Femoral artery and vein in femoral sheath

Fig.15.4
The inguinal canal.
(a) The superficial inguinal ring. The external spermatic fascia has been removed
(b) After removal of the external oblique

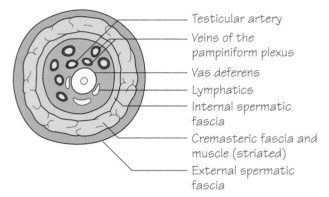

Testicular artery

Veins of the pampiniform plexus

Vas deferens

Lymphatics

Internal spermatic fascia

Cremasteric fascia and muscle (striated)

External spermatic fascia

Fig.15.5
A schematic cross-section through the spermatic cord

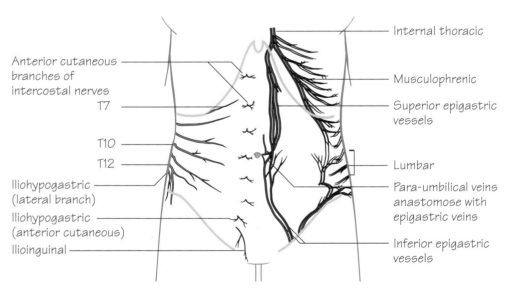

Internal thoracic

Musculophrenic

Superior epigastric vessels

Lumbar

Para-umbilical veins anastomose with epigastric veins

Inferior epigastric vessels

Anterior cutaneous branches of intercostal nerves

T7

T10

T12

Iliohypogastric (lateral branch)

Iliohypogastric (anterior cutaneous)

Ilioinguinal

Fig.15.6
The nerves and vessels of the abdominal wall

The anterior abdominal wall comprises the: skin, superficial fascia, abdominal muscles (and their respective aponeuroses), transversalis fascia, extraperitoneal fat and parietal peritoneum.

Skin (Fig. 15.6)
The skin of the abdominal wall is innervated by the anterior rami of the lower six thoracic intercostal and iliohypogastric (L1) nerves.

Fascia (Fig. 15.2)
There is no deep fascia in the trunk. The superficial fascia is composed of two layers:
- A superficial fatty layer – *Camper's fascia* – which is continuous with the superficial fat over the rest of the body.
- A deep fibrous (membranous) layer – *Scarpa's fascia* – which fades above and laterally but blends below with the fascia lata of the thigh, just below the inguinal ligament and extends into: the penis as a tubular sheath; the wall of the scrotum and the perineum posteriorly where it is referred to as Colle's fascia. At this site, it is fused with the perineal body and posterior margin of the perineal membrane, as well as with the pubic arch laterally.

Muscles of the anterior abdominal wall (Fig. 15.1 and 15.3)
These comprise: external oblique, internal oblique, transversus abdominis, rectus abdominis and pyramidalis (see Muscle index, p. 178).

As in the intercostal space, the neurovascular structures pass in the neurovascular plane between the internal oblique and transversus muscle layers.

The rectus sheath (Fig. 15.3)
The rectus sheath encloses the rectus abdominis muscles. It also contains the superior and inferior epigastric vessels and anterior rami of the lower six thoracic nerves.

The sheath is made up from the aponeuroses of the muscles of the anterior abdominal wall. The *linea alba* represents the fusion of the aponeuroses in the midline. Throughout the major part of the length of the rectus, the aponeuroses of the external oblique and the anterior layer of the internal oblique lie in front of the muscle and the posterior layer of the internal oblique and transversus behind. The composition of the sheath is, however, different above the costal margin and above the pubic symphysis:
- **Above the costal margin**: only the external oblique aponeurosis is present and forms the anterior sheath.
- **Above the pubic symphysis**: about halfway between the umbilicus and pubic symphysis, the layers passing behind the rectus muscle gradually fade out and, from this point, all aponeuroses pass anterior to the rectus muscle, leaving only the transversalis fascia posteriorly.

The lateral border of the rectus – the *linea semilunaris* – can usually be identified in thin subjects. It crosses the costal margin in the transpyloric plane.

Three tendinous intersections firmly attach the anterior sheath wall to the muscle itself. They are situated at the level of the xiphoid, the umbilicus, and one between these two. These give the abdominal the 'six-pack' appearance in muscular individuals.

Arteries of the abdominal wall (Fig. 15.6)
These include the *superior* and *inferior epigastric arteries* (branches of the internal thoracic and external iliac arteries, respectively) and the *deep circumflex iliac artery* (a branch of the external iliac artery) anteriorly. The two lower *intercostal* and four *lumbar* arteries supply the wall posterolaterally.

Veins of the abdominal wall (Fig. 15.6)
The abdominal wall is a site of porto-systemic anastomosis. The *lateral thoracic*, *lumbar* and *superficial epigastric* tributaries of the systemic circulation anastomose around the umbilicus with the *para-umbilical* veins which accompany the ligamentum teres and drain into the portal circulation.

Lymph drainage of the abdominal wall
See section 'The lymphatic drainage of the abdomen and pelvis' in Chapter 17.

The inguinal canal (Fig. 15.4)
The canal is approximately 4 cm long and allows the passage of the spermatic cord (round ligament in the female) through the lower abdominal wall. It passes obliquely from the *deep inguinal ring* in a medial direction to the *superficial inguinal ring*.
- **The deep ring**: is an opening in the transversalis fascia and lies halfway between the anterior superior iliac spine and the pubic tubercle. The inferior epigastric vessels pass medial to the deep ring.
- **The superficial ring**: is not a ring but a triangular-shaped defect in the external oblique aponeurosis lying above and medial to the pubic tubercle.

The walls of the inguinal canal (Fig. 15.4)
- **Anterior wall**: the external oblique covers the length of the canal anteriorly. It is reinforced in its lateral third by the internal oblique.
- **Superior wall**: the internal oblique arches posteriorly to form the roof of the canal.
- **Posterior wall**: the transversalis fascia forms the lateral part of the posterior wall. The conjoint tendon (the combined common insertion of the internal oblique and transversus into the pectineal line) forms the medial part of the posterior wall.
- **Inferior wall**: the inguinal ligament.

Contents of the inguinal canal
- The spermatic cord (or round ligament in the female).
- The ilioinguinal nerve (L1).

The spermatic cord (Fig. 15.5)
The spermatic cord is covered by three layers which arise from the layers of the lower abdominal wall as the cord passes through the inguinal canal. These are the:
- **External spermatic fascia** from the external oblique aponeurosis.
- **Cremasteric fascia and muscle** from the internal oblique aponeurosis.
- **Internal spermatic fascia** from the transversalis fascia.
  The contents of the spermatic cord include the:
- **Ductus (vas) deferens (or round ligament)**.
- **Testicular artery** which is a branch of the abdominal aorta.
- **Pampiniform plexus of veins** which coalesce to form the testicular vein in the region of the deep ring.
- **Lymphatics** from the testis and epididymis draining to the pre-aortic nodes.
- **Autonomic nerves**.

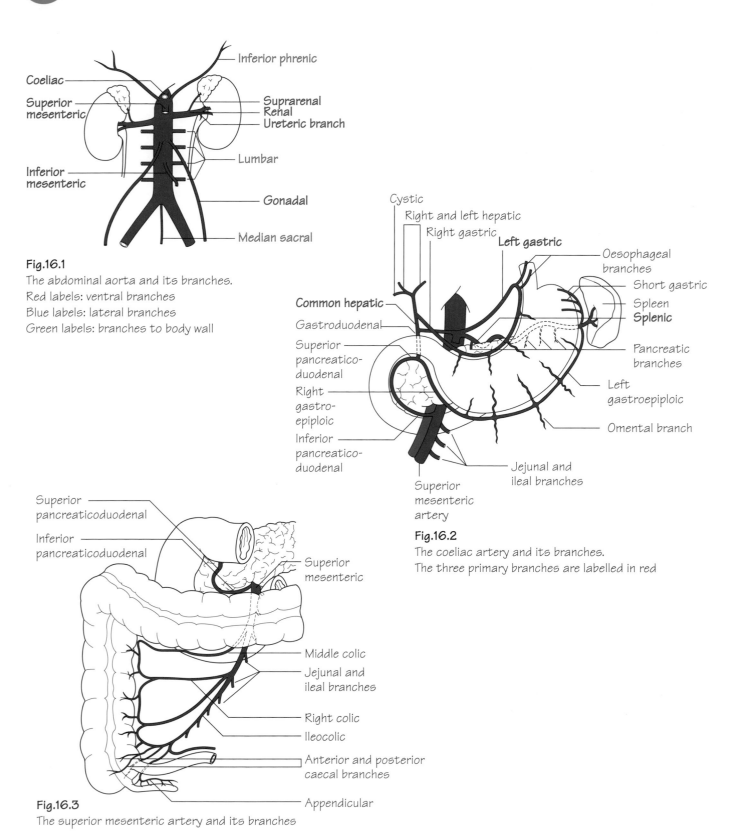

Fig.16.1

The abdominal aorta and its branches.
Red labels: ventral branches
Blue labels: lateral branches
Green labels: branches to body wall

Fig.16.2

The coeliac artery and its branches.
The three primary branches are labelled in red

Fig.16.3

The superior mesenteric artery and its branches

Anatomy at a Glance, Third Edition. Omar Faiz, Simon Blackburn and David Moffat.
© 2011 Blackwell Publishing Ltd. Published 2011 by Blackwell Publishing Ltd.

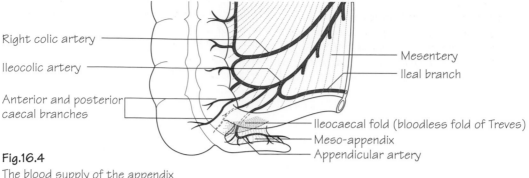

Right colic artery

Ileocolic artery

Anterior and posterior
caecal branches

Mesentery

Ileal branch

Ileocaecal fold (bloodless fold of Treves)

Meso-appendix

Appendicular artery

Fig.16.4
The blood supply of the appendix

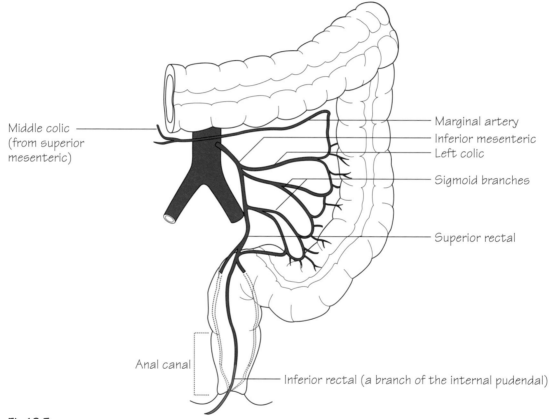

Middle colic
(from superior
mesenteric)

Marginal artery

Inferior mesenteric

Left colic

Sigmoid branches

Superior rectal

Anal canal

Inferior rectal (a branch of the internal pudendal)

Fig.16.5
The inferior mesenteric artery and its branches.
Note the anastomosis with the inferior rectal artery (green) halfway down the anal canal

The abdominal aorta (Fig. 16.1)

The abdominal aorta is a continuation of the thoracic aorta as it passes under the median arcuate ligament of the diaphragm. It descends in the retroperitoneum and ultimately bifurcates into left and right common iliac arteries, to the left of the midline at the level of L4. The vertebral bodies and intervertebral discs lie behind the aorta, whilst anteriorly, from above downwards, lie the coeliac trunk and its branches, the lesser sac, the body of the pancreas, the third part of the duodenum and the parietal peritoneum. The main relation to the right of the abdominal aorta is the inferior vena cava, whilst to the left lie the duodenojejunal junction and inferior mesenteric vein.

The main abdominal branches of the abdominal aorta include the following:
- **Coeliac trunk**: supplies the embryonic foregut from the lower third of the oesophagus to the second part of the duodenum.
- **Superior mesenteric artery**: supplies the midgut from the second part of the duodenum to the distal transverse colon.
- **Renal arteries**.
- **Gonadal arteries**.
- **Inferior mesenteric artery**: supplies the hindgut from the distal transverse colon to the upper half of the anal canal.

The coeliac trunk (Fig. 16.2)

This trunk arises from the aorta at the level of T12/L1 and, after a short course, divides into three terminal branches:
- **Left gastric artery**: passes upwards to supply the lower oesophagus by branches which ascend through the oesophageal hiatus in the diaphragm. The left gastric artery itself, then, descends in the lesser omentum along the lesser curve of the stomach, which it supplies.
- **Splenic artery**: passes along the superior border of the pancreas in the posterior wall of the lesser sac to reach the upper pole of the left kidney. From here, it passes to the hilum of the spleen in the lienorenal ligament. The splenic artery also gives rise to *short gastric branches*, which supply the stomach fundus, and a *left gastroepiploic branch*, which passes in the gastrosplenic ligament to reach and supply the greater curve of the stomach.
- **Common hepatic artery**: descends to the right towards the first part of the duodenum in the posterior wall of the lesser sac. It, then, passes between the layers of the free border of the lesser omentum which conveys it to the porta hepatis in close relation to the portal vein and bile duct (these structures together constitute the anterior margin of the epiploic foramen). Before reaching the porta hepatis it divides into *right* and *left hepatic arteries* and from the right branch the *cystic artery* is usually given off. Prior to its ascent towards the porta hepatis the hepatic artery gives rise to *gastroduodenal* and *right gastric branches*. The latter passes along the lesser curve of the stomach to supply it. The former passes behind the first part of the duodenum and then branches further into *superior pancreaticoduodenal* and *right gastroepiploic branches*. The right gastroepiploic branch runs along the lower part of the greater curvature to supply the stomach.

The superior mesenteric artery (Fig. 16.3)

The superior mesenteric artery arises from the abdominal aorta at the level of L1. From above downwards, it passes over the left renal vein behind the neck of the pancreas, over the uncinate process of the pancreas and anterior to the third part of the duodenum. It then passes obliquely downwards and towards the right iliac fossa between the layers of the mesentery of the small intestine where it divides into its terminal branches. The branches of the superior mesenteric artery include the following:
- **Inferior pancreaticoduodenal artery**: supplies the lower half of the duodenum and pancreatic head.
- **Ileocolic artery**: passes in the root of the mesentery over the right ureter and gonadal vessels to reach the caecum where it divides into terminal *caecal* and *appendicular branches* (Fig. 16.4).
- **Jejunal and ileal branches**: a total of 12–15 branches arises from the left side of the artery. These branches divide and reunite within the small bowel mesentery to form a series of arcades which, then, give rise to small straight terminal branches which supply the gut wall.
- **Right colic artery**: passes horizontally in the posterior abdominal wall to supply the ascending colon.
- **Middle colic artery**: courses in the transverse mesocolon to supply the proximal two-thirds of the transverse colon.

The renal arteries

These arise from the abdominal aorta at the level of L2.

The gonadal arteries (ovarian or testicular)

These arteries arise from below the renal arteries and descend obliquely on the posterior abdominal wall to reach the ovary in the female, or pass through the inguinal canal in the male to reach the testis.

The inferior mesenteric artery (Fig. 16.5)

The inferior mesenteric artery arises from the abdominal aorta at the level of L3. It passes downwards and to the left and crosses the left common iliac artery, where it changes its name to the *superior rectal artery*. Its branches include:
- **The left colic artery**: supplies the distal transverse colon, the splenic flexure and upper descending colon.
- **Two or three sigmoid branches**: pass into the sigmoid mesocolon and supply the lower descending and sigmoid colon.
- **The superior rectal artery**: passes into the pelvis behind the rectum to form an anastomosis with the middle and inferior rectal arteries. It supplies the rectum and upper half of the anal canal.

The *marginal artery (of Drummond)* is an anastomosis of the colic arteries at the margin of the large intestine. This establishes a strong collateral circulation throughout the colon.

Clinical notes

- **Abdominal aortic aneurysm**: atheromatous degeneration of the aorta can result in progressive aortic dilatation. Aneurysms of the abdominal aorta most commonly occur below the level of the renal arteries. The majority are symptomless and consequently expand undetected. Once the diameter of the aorta exceeds 5 cm, there is a significant risk of rupture. Only half of the patients that suffer ruptured abdominal aortic aneurysms survive to reach hospital. Of these patients, approximately half survive the emergency operation.

17 The veins and lymphatics of the abdomen

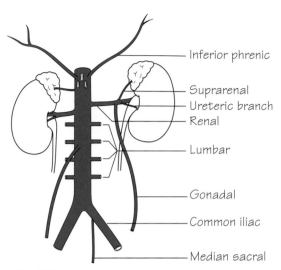

Fig.17.1
The inferior vena cava and its tributaries

- Inferior phrenic
- Suprarenal
- Ureteric branch
- Renal
- Lumbar
- Gonadal
- Common iliac
- Median sacral

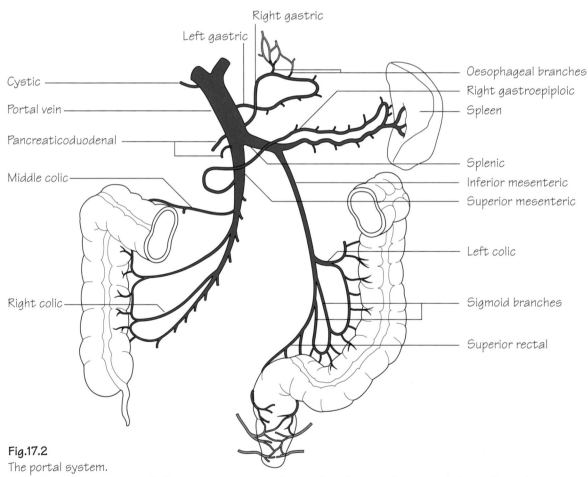

Fig.17.2
The portal system.
Note the anastomoses with the systemic system (orange) in the oesophagus and the anal canal

- Right gastric
- Left gastric
- Cystic
- Portal vein
- Pancreaticoduodenal
- Middle colic
- Right colic
- Oesophageal branches
- Right gastroepiploic
- Spleen
- Splenic
- Inferior mesenteric
- Superior mesenteric
- Left colic
- Sigmoid branches
- Superior rectal

The portal vein (Fig. 17.2)

The portal venous system receives blood from the length of the gut between the lower third of the oesophagus and the upper half of the anal canal as well as from the spleen, pancreas and gall bladder. It serves to transfer blood to the liver, where the products of digestion are metabolised and stored. Blood from the liver ultimately drains into the inferior vena cava through the hepatic veins. The portal vein is formed behind the neck of the pancreas by the union of the superior mesenteric and splenic veins. It then passes behind the first part of the duodenum in front of the inferior vena cava. The portal vein ascends towards the porta hepatis as the anterior margin of the epiploic foramen (of Winslow) in the lesser omentum. At the porta hepatis, it divides into right and left branches. The veins that correspond to the branches of the coeliac and superior mesenteric arteries drain into the portal vein or one of its tributaries. The inferior mesenteric vein drains into the splenic vein adjacent to the fourth part of the duodenum.

Porto-systemic anastomoses

A number of connections occur between the portal and systemic circulations. When the direct pathway through the liver becomes congested (such as in cirrhosis), the pressure within the portal vein rises and, under these circumstances, the porto-systemic anastomoses form an alternative route for the blood to take. The sites of porto-systemic anastomosis include:

• **The lower oesophagus (p. 19)**: formed by tributaries of the left gastric (portal) and oesophageal (systemic) veins.
• **The anal canal**: formed by the superior rectal (portal) and middle and inferior rectal (systemic) veins.
• **The bare area of the liver**: formed by the small veins of the portal system and the phrenic veins (systemic).
• **The periumbilical region**: formed by small para-umbilical veins which ultimately drain into the portal vein and the superficial veins of the anterior abdominal wall (systemic).

The inferior vena cava (Fig. 17.1)

The inferior vena cava is formed by the union of the common iliac veins in front of the body of L5. It ascends in the retroperitoneum on the right side of the abdominal aorta. As it passes superiorly, it forms the posterior wall of the epiploic foramen (of Winslow) and is embedded in the bare area of the liver, in front of the right suprarenal gland. The inferior vena cava passes through the caval opening in the diaphragm at the level of T8 and drains into the right atrium.

The lymphatic drainage of the abdomen and pelvis

The abdominal wall

Lymph from the skin of the anterolateral abdominal wall, above the level of the umbilicus, drains to the anterior axillary lymph nodes. Efferent lymph from the skin, below the umbilicus, drains to the superficial inguinal nodes.

The lymph nodes and trunks

The two main lymph node groups of the abdomen are closely related to the aorta. These comprise the pre-aortic and para-aortic groups.

• The *pre-aortic nodes* are arranged around the three ventral branches of the aorta and, consequently, receive lymph from the territories that are supplied by these branches. This includes most of the gastrointestinal tract, liver, gall bladder, spleen and pancreas. The efferent vessels from the pre-aortic nodes coalesce to form a variable number of *intestinal trunks* which deliver the lymph to the cisterna chyli.
• The *para-aortic nodes* are arranged around the lateral branches of the aorta, and drain lymph from their corresponding territories, i.e. the kidneys, adrenals, gonads and abdominal wall, as well as the common iliac nodes. The efferent vessels from the para-aortic nodes coalesce to form a variable number of *lumbar trunks* which deliver the lymph to the cisterna chyli.

Cisterna chyli

The cisterna chyli is a lymphatic sac that lies anterior to the bodies of the 1st and 2nd lumbar vertebrae. It is formed by the confluence of the intestinal trunks, the lumbar trunks and lymphatics from the lower thoracic wall. It serves as a receptacle for lymph from the abdomen and lower limbs, which is then relayed to the thorax by the thoracic duct (p. 19).

The lymphatic drainage of the stomach

All lymph from the stomach ultimately drains to the coeliac nodes. For the purposes of description, the stomach can be divided into four quarters where lymph drains to the nearest appropriate group of local nodes, which are associated with the draining veins.

The lymphatic drainage of the testes

Lymph from the skin of the scrotum and the tunica albuginea drains to the superficial inguinal nodes. Lymph from the testes, however, drains along the course of the testicular veins to the para-aortic nodes. A malignancy of the scrotal skin might, therefore, result in palpable enlargement of the superficial inguinal nodes, whereas testicular tumours metastasise to the para-aortic nodes.

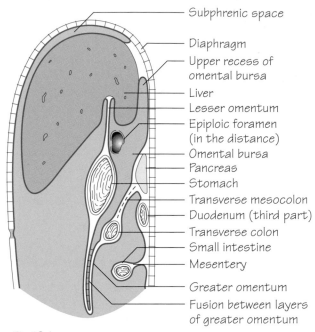

Subphrenic space
Diaphragm
Upper recess of omental bursa
Liver
Lesser omentum
Epiploic foramen (in the distance)
Omental bursa
Pancreas
Stomach
Transverse mesocolon
Duodenum (third part)
Transverse colon
Small intestine
Mesentery
Greater omentum
Fusion between layers of greater omentum

Fig.18.1
A vertical section through the abdomen to show the peritoneal relations.
Lesser sac ▭
Greater sac ▭

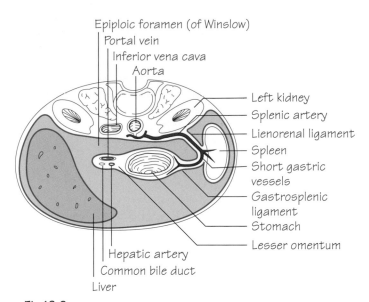

Epiploic foramen (of Winslow)
Portal vein
Inferior vena cava
Aorta
Left kidney
Splenic artery
Lienorenal ligament
Spleen
Short gastric vessels
Gastrosplenic ligament
Stomach
Lesser omentum
Hepatic artery
Common bile duct
Liver

Fig.18.2
A horizontal section through the abdomen.
Note how the epiploic foramen lies between two major veins

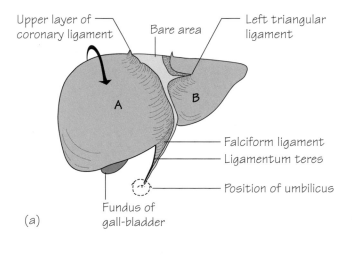

Upper layer of coronary ligament
Bare area
Left triangular ligament
A
B
Falciform ligament
Ligamentum teres
Position of umbilicus
Fundus of gall-bladder
(a)

Fig.18.3
The peritoneal relations of the liver
(a) Seen from in front
(b) The same liver rotated in the direction of the arrow to show the upper and posterior surfaces.
The narrow spaces between the liver and the diaphragm labelled A and B are the right and left subphrenic spaces

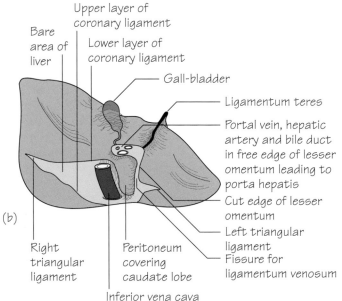

Upper layer of coronary ligament
Bare area of liver
Lower layer of coronary ligament
Gall-bladder
Ligamentum teres
Portal vein, hepatic artery and bile duct in free edge of lesser omentum leading to porta hepatis
Cut edge of lesser omentum
Left triangular ligament
Fissure for ligamentum venosum
Right triangular ligament
Peritoneum covering caudate lobe
Inferior vena cava
(b)

The mesenteries and layers of the peritoneum

The transverse colon, stomach, spleen and liver, each have attached to them *two* 'mesenteries' – double layers of peritoneum containing arteries and their accompanying veins, nerves and lymphatics – while the small intestine and sigmoid colon have only one. All the other viscera are retroperitoneal. The mesenteries and their associated arteries are as follows:

- **The colon (Fig. 18.1)**: (1) The *transverse mesocolon* (the middle colic artery). (2) The posterior two layers of the *greater omentum*.
- **The stomach (Fig. 18.1)**: (1) The *lesser omentum* (the left and right gastric arteries and, in its free border, the hepatic artery, portal vein and bile duct). (2) The anterior two layers of the *greater omentum* (the right and left gastroepiploic arteries and their omental branches).
- **The spleen (Fig. 18.2)**: (1) The *lienorenal ligament* (the splenic artery). (2) The *gastrosplenic ligament* (the short gastric and left gastroepiploic arteries).
- **The liver (Fig. 18.3)**: (1) The *falciform ligament* and the two layers of the *coronary ligament* with their sharp edges, the *left* and *right triangular ligaments*. This mesentery is exceptional in that the layers of the coronary ligament are widely separated so that the liver has a *bare area* directly in contact with the diaphragm (the obliterated umbilical vein in the free edge of the falciform ligament and numerous small veins in the bare area, p. 47). (2) The *lesser omentum* (already described).
- **The small intestine (Fig. 18.1)**: (1) The *mesentery of the small intestine* (the superior mesenteric artery and its branches).
- **The sigmoid colon**: (1) The *sigmoid mesocolon* (the sigmoid arteries and their branches).

The peritoneal cavity (Figs. 18.1 and 18.2)

- The course of the peritoneum may best be considered by starting at the root of the *transverse mesocolon*. The two layers of this mesentery are attached to the anterior surface of the pancreas, the second part of the duodenum and the front of the left kidney. They envelop the transverse colon and continue downwards to form the posterior two layers of the *greater omentum*, which hangs down over the coils of the small intestine. They, then, turn back on themselves to form the anterior two layers of the omentum, and reach the greater curvature of the stomach. The four layers of the greater omentum are fused and impregnated with fat. The greater omentum plays an important role in limiting the spread of infection in the peritoneal cavity.
- From its attachment to the pancreas, the lower layer of the transverse mesocolon turns downwards to become the parietal peritoneum of the posterior abdominal wall from which it is reflected to form the *mesentery of the small intestine* and the *sigmoid mesocolon*.
- The upper layer of the transverse mesocolon passes upwards to form the parietal peritoneum of the posterior abdominal wall, covering the upper part of the pancreas, the left kidney and its suprarenal, the aorta and the origin of the coeliac artery (the '*stomach bed*'). It, thus, forms the posterior wall of the omental bursa. It then covers the diaphragm and continues onto the anterior abdominal wall.
- From the diaphragm and anterior abdominal wall it is reflected onto the liver to form its 'mesentery' in the form of the two layers of the *falci-form ligament*. At the liver, the left layer of the falciform ligament folds back on itself to form the sharp edge of the *left triangular ligament*, while the right layer turns back on itself to form the upper and lower layers of the *coronary ligament* with its sharp-edged *right triangular ligament*. The layers of the coronary ligament are widely separated, so that a large area of liver between them—the *bare area*—is directly in contact with the diaphragm. The inferior vena cava is embedded in the bare area (Fig. 18.3).
- From the undersurface of the liver another 'mesentery' passes from the fissure for the ligamentum venosum to the lesser curvature of the stomach to form the *lesser omentum*.
- The *lesser omentum* splits to enclose the stomach and is continuous with the two layers of the *greater omentum* already described. The lesser omentum has a right free border which contains the portal vein, the hepatic artery and the common bile duct.
- In the region of the spleen there are two more 'mesenteries' which are continuous with the lesser and greater omenta. These are the *lienorenal ligament*, a double layer of peritoneum reflected from the front of the left kidney to the hilum of the spleen, and the *gastrosplenic ligament*, which passes from the hilum of the spleen to the greater curvature of the stomach (Fig. 18.2).
- The *mesentery of the small intestine* is attached to the posterior abdominal wall from the duodenojejunal flexure to the ileocolic junction.
- The *sigmoid mesocolon* passes from a V-shaped attachment on the posterior abdominal wall to the sigmoid colon.
- The general peritoneal cavity comprises the main cavity – the *greater sac* – and a diverticulum from it – the *omental bursa* (*lesser sac*). The omental bursa lies between the stomach and the stomach bed to allow free movement of the stomach. It lies behind the stomach, the lesser omentum and the caudate lobe of the liver and in front of the structures of the stomach bed. The left border is formed by the hilum of the spleen and the lienorenal and gastrosplenic ligaments.
- The communication between the greater and lesser sacs is the *epiploic foramen* (*foramen of Winslow*). It lies behind the free border of the lesser omentum and its contained structures, below the caudate process of the liver, in front of the inferior vena cava and above the first part of the duodenum.
- The *subphrenic spaces* are the parts of the greater sac that lie between the diaphragm and the upper surface of the liver. The right and left spaces are separated by the falciform ligament.
- In the pelvis, the parietal peritoneum covers the upper two-thirds of the rectum whence it is reflected, in the female, onto the posterior fornix of the vagina and the back of the uterus to form the *recto-uterine pouch* (*pouch of Douglas*). In the male, it passes onto the back of the bladder to form the *rectovesical pouch*.

The anterior abdominal wall

- The peritoneum of the deep surface of the anterior abdominal wall shows a central ridge from the apex of the bladder to the umbilicus produced by the *median umbilical ligament*. This is the embryological remnant of the urachus. Two *medial umbilical ligaments* converge to the umbilicus from the pelvis. They represent the obliterated umbilical arteries of the fetus. The *ligamentum teres* is a fibrous band in the free margin of the falciform ligament. It represents the obliterated left umbilical vein.

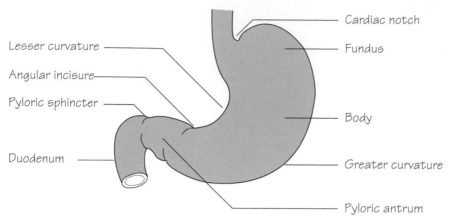

Fig.19.1
The subdivisions of the stomach

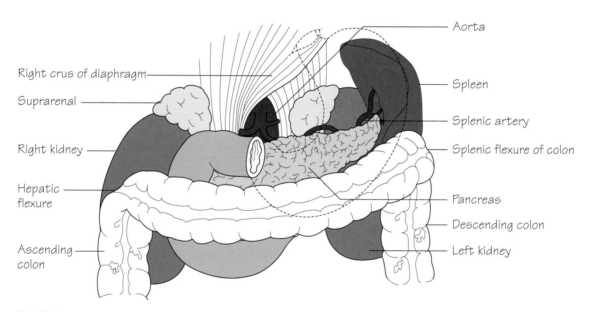

Fig.19.2
The stomach bed. For more detail see fig.23.1.
The stomach is outlined but the shape is by no means constant

The embryonic gut is divided into the foregut, midgut and hindgut, supplied by the coeliac, superior mesenteric and inferior mesenteric arteries, respectively. The foregut extends from the oesophagus to the entrance of the common bile duct into the second part of the duodenum. The midgut extends down to two-thirds of the way along the transverse colon. It largely develops outside the abdomen until this congenital 'umbilical hernia' is reduced during the 8th–10th week of gestation (see Chapter 31). The hindgut extends down to include the upper half of the anal canal.

The abdominal oesophagus

- The abdominal oesophagus measures approximately 1 cm in length.
- It is accompanied by the anterior and posterior vagal trunks from the left and right vagi and the oesophageal branches of the left gastric artery.
- The lower third of the oesophagus is a site of porto-systemic venous anastomosis. This is formed between tributaries of the left gastric and azygos veins (p. 19).

The stomach (Figs. 19.1, 19.2, and 20.3)

- The notch on the lesser curve, at the junction of the body and pyloric antrum, is the *incisura angularis*.
- The *pyloric sphincter* controls the release of stomach contents into the duodenum. The sphincter is composed of a thickened layer of circular smooth muscle which acts as an anatomical, as well as physiological, sphincter. The junction of the pylorus and duodenum can be seen externally as a constriction with an overlying vein—the *prepyloric vein (of Mayo)*.
- The *cardiac orifice* represents the point of entry for oesophageal contents into the stomach. The cardiac sphincter acts to prevent reflux of stomach contents into the oesophagus. Unlike the pylorus, there is no discrete anatomical sphincter at the cardia; however, multiple factors contribute towards its mechanism, preventing reflux of gastric content into the oesophagus. These include: the arrangement of muscle fibres at the cardiac orifice acting as a physiological sphincter, the angle at which the oesophagus enters the stomach producing a valve effect, the right crus of the diaphragm surrounding the oesophagus and compression of the short segment of intra-abdominal oesophagus by increased intra-abdominal pressure during straining.
- The *lesser omentum* is attached to the lesser curvature and the *greater omentum* to the greater curvature of the stomach. The omenta contain the blood and lymphatic supply to the stomach.
- The mucosa of the stomach is thrown into folds – *rugae*.
- *Blood supply* (see Fig. 16.2): the arterial supply to the stomach is exclusively from branches of the coeliac axis; venous drainage is to the portal system (see Fig. 17.2).
- *Nerve supply*: the anterior and posterior vagal trunks arise from the oesophageal plexuses and enter the abdomen through the oesophageal hiatus. The *hepatic* branches of the anterior vagus pass to the liver. The *coeliac* branch of the posterior vagus passes to the coeliac ganglion from where it proceeds to supply the intestine down to the distal transverse colon. The anterior and posterior vagal trunks descend along the lesser curve as the anterior and posterior nerves of *Latarjet* from which terminal branches arise to supply the stomach. The vagi provide a motor and secretory supply to the stomach. The latter includes a supply to the acid-secreting part – the *body*.

The duodenum (Figs. 20.3, 23.1, and 23.2)

The duodenum is the first part of the small intestine. It is approximately 25 cm long and curves around the head of the pancreas. Its primary function is in the absorption of digested products. Despite its relatively short length, the surface area is greatly enhanced by the mucosa being thrown into folds bearing villi, which are visible only at a microscopic level. With the exception of the first 2.5 cm, the duodenum is a retroperitoneal structure. It is studied in four parts:

- **First part (5 cm)**.
- **Second part (7.5 cm)**: this part descends around the head of the pancreas. Internally, in the mid-section, a small eminence may be found on the posteromedial aspect of the mucosa—the *duodenal papilla*. This structure represents the site of the common opening of the bile duct and *main pancreatic duct (of Wirsung)*. The *sphincter of Oddi* guards this common opening. A smaller *subsidiary pancreatic duct (of Santorini)* opens into the duodenum at a small distance above the papilla.
- **Third part (10 cm)**: this part is crossed anteriorly by the root of the mesentery and superior mesenteric vessels.
- **Fourth part (2.5 cm)**: this part terminates as the duodenojejunal junction. The termination of the duodenum is demarcated by a peritoneal fold stretching from the junction to the right crus of the diaphragm covering the *suspensory ligament of Treitz*. The terminal part of the inferior mesenteric vein lies adjacent to the duodenojejunal junction and serves as a useful landmark.

Blood supply (see Fig. 16.2): the superior and inferior pancreaticoduodenal arteries supply the duodenum and run between this structure and the pancreatic head. The superior artery arises from the coeliac axis and the inferior from the superior mesenteric artery.

Clinical notes

- **Peptic ulcer disease**: most peptic ulcers occur in the stomach and proximal duodenum. They arise as a result of an imbalance between acid secretion and mucosal defences. *Helicobacter pylori* infection is a significant aetiological factor, and the eradication of this organism and the attenuation of acid secretion form the cornerstones of medical treatment. Today, only a minority of elective cases require surgery. 'Very highly selective vagotomy' is a low-risk technique in which the afferent vagal fibres to the acid-secreting gastric body are denervated, thus not compromising the motor supply to the stomach and, hence, bypassing the need for a simultaneous drainage procedure (e.g. gastrojejunostomy). The complications of untreated duodenal ulcers include bleeding and perforation. Classically, ulcers occurring in the posterior duodenal bulb cause bleeding as a result of erosion into the gastroduodenal artery. In contrast, ulcers arising in the anterior duodenal bulb tend to perforate into the peritoneal cavity.

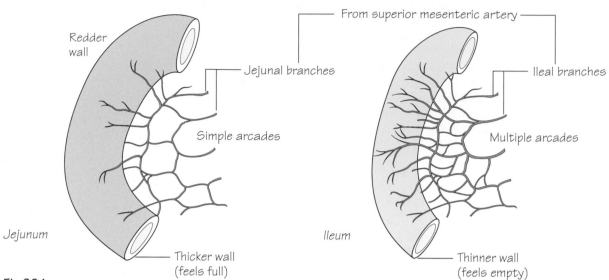

Fig.20.1

The jejunum and the ileum can be distinguished by their colour, feel and the complexity of the arterial arcades

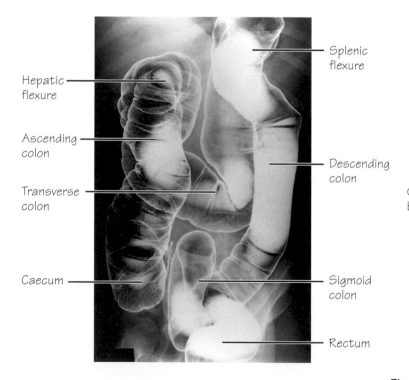

Fig.20.2

Normal double contrast barium enema showing the anatomy of the colon (see p.54). This Investigation is performed by instilling contrast rectally, followed by air.

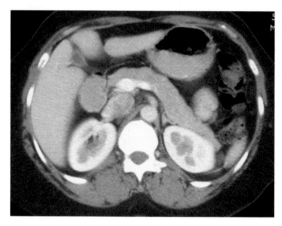

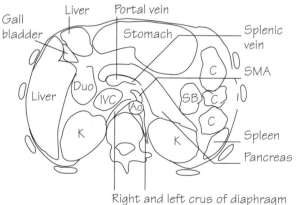

Fig.20.3

CT scan of the abdomen at the level of the transpyloric plane. The diagram identifies the structures present. K–kidney, Ao–aorta, IVC–inferior vena cava, SB–bowel, C–colon, SMA–superior mesenteric artery

The small intestine (Fig. 20.1)

The small intestine is approximately 4–6 m long and comprises the duodenum, jejunum and ileum. A large internal surface area throughout the small intestine facilitates absorption of digested products. The small intestine is suspended from the posterior abdominal wall by its mesentery which contains the superior mesenteric vessels, lymphatics and autonomic nerves. The origin of the mesentery measures approximately 15 cm and passes from the duodenojejunal flexure to the right sacro-iliac joint. The distal border is obviously of the same length as the intestine. No sharp distinction occurs between the jejunum and ileum; however, certain characteristics help to distinguish between them:

• Excluding the duodenum, the proximal two-fifths of the small intestine comprises the jejunum, whereas the remaining distal three-fifths comprises the ileum. Loops of jejunum tend to occupy the umbilical region, whereas the ileum occupies the lower abdomen and pelvis.

• The mucosa of the small intestine is thrown into circular folds – the *plicae circulares*. These are more prominent in the jejunum than in the ileum.

• The diameter of the jejunum tends to be greater than that of the ileum.

• The mesentery to the jejunum tends to be thicker than that to the ileum.

• The superior mesenteric vessels (see Figs. 16.3 and 20.1) pass over the third part of the duodenum to enter the root of the mesentery and pass towards the right iliac region on the posterior abdominal wall. Jejunal and ileal branches arise and divide and reanastomose within the mesentery to produce arcades. End-artery vessels arise from the arcades to supply the gut wall. The arterial supply to the jejunum consists of few arcades and little terminal branching, whereas the vessels to the ileum form numerous arcades and much terminal branching of end-arteries passing to the gut wall.

Clinical notes

• **Small bowel obstruction**: small bowel obstruction (SBO) can occur due to luminal, mural or extraluminal factors that result in blockage. Post-surgical adhesions and herniae are the most frequent causes. Many cases of adhesional obstruction resolve with conservative measures only. If any deterioration in the clinical picture occurs, indicating intestinal infarction or perforation, an exploratory laparotomy is, however, mandatory. The classical X-ray features of SBO are those of dilated small bowel loops. The latter can be distinguished from large bowel loops as the *valvulae conniventes* (radiographic representations of the plicae circulares) can be identified traversing the entire lumen. In contrast, the large bowel haustra traverse the lumen only partially (Fig. 20.2).

The lower gastrointestinal tract

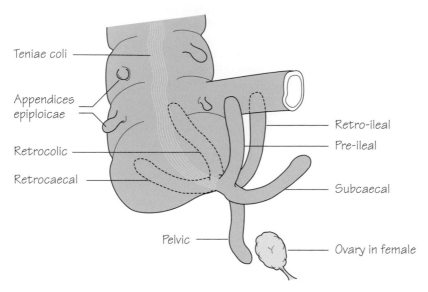

Fig.21.1

The various positions in which the appendix may be found.
In the pelvic position the appendix may be close to the ovary in the female

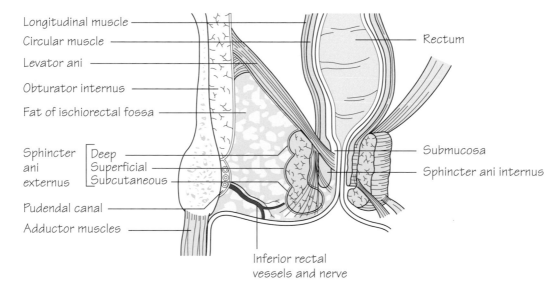

Fig.21.2

A coronal section through the pelvis to show the
anal sphincters and the ischiorectal fossa

The caecum and colon (Figs. 16.3, 16.5, 20.2 and 21.1)

In adults, the large bowel measures approximately 1.5 m. The blood supply to the colon and rectum is described in Chapter 16 (p. 44). The caecum (ascending, transverse and descending) and sigmoid colon have similar characteristic features, which include:

- **Appendices epiploicae (Fig. 21.1):** these are fat-laden peritoneal tags present over the surface of the caecum and colon.
- **Teniae coli (Fig. 21.1):** these are three flattened bands representing the condensed longitudinal muscular coat of the large intestine. They course from the base of the appendix (and form a useful way of locating this structure at surgery) to the recto-sigmoid junction.

- **Sacculations**: because the teniae are shorter than the bowel itself, the colon takes on a sacculated appearance. These sacculations are visible not only at operation but also radiographically. On a plain abdominal X-ray, the colon, which appears radiotranslucent because of the gas within, has shelf-like processes (*haustra*) which partially project into the lumen.

The transverse and sigmoid colon are each attached to the posterior abdominal wall by their respective mesocolons and are covered entirely by peritoneum. Conversely, the ascending and descending colon normally possess no mesocolon. They are adherent to the posterior abdominal wall and covered only anteriorly by peritoneum.

The appendix (Fig. 21.1)

The appendix varies enormously in length but, in adults, is approximately 5–15 cm long. The base of the appendix arises from the posteromedial aspect of the caecum; however, the lie of the appendix itself is highly variable. The surface marking for the base of the appendix is represented by *McBurney's* point (see p. 65). In most cases, the appendix lies in the retrocaecal position but other positions frequently occur. The appendix has the following characteristic features:

- It has a small mesentery that descends behind the terminal ileum. The only blood supply to the appendix, the appendicular artery (a branch of the ileocolic), courses within its mesentery (see Fig. 16.4). In cases of appendicitis, the appendicular artery ultimately thromboses. When this occurs, gangrene and perforation of the appendix inevitably supervene.
- The appendix has a lumen which is relatively wide in infants and gradually narrows throughout life, often becoming obliterated in the elderly.
- The teniae coli of the caecum lead to the base of the appendix.
- The *bloodless fold of Treves* (ileocaecal fold) is the name given to a small peritoneal reflection passing from the anterior terminal ileum to the appendix. Despite its name, it is not an avascular structure!

The rectum (Figs. 16.5, 20.2 and 21.2)

- The rectum measures 10–15 cm in length. It commences in front of the 3rd sacral vertebra as a continuation of the sigmoid colon and follows the curve of the sacrum anteriorly. It turns backwards abruptly in front of the coccyx to become the anal canal.
- The mucosa of the rectum is thrown into three horizontal folds that project into the lumen – the *valves of Houston*.
- The rectum lacks haustrations. The teniae coli fan out over the rectum to form anterior and posterior bands.
- The rectum is slightly dilated at its lower end – the *ampulla*, and is supported laterally by the levator ani.
- Peritoneum covers the upper two-thirds of the rectum anteriorly but only the upper third laterally. In the female, it is reflected forwards onto the uterus forming the *recto-uterine pouch* (*pouch of Douglas*). The rectum is separated from anterior structures by a tough fascial sheet – the *rectovesical (Denonvilliers) fascia*.

The anal canal (Fig. 21.2)

The anorectal junction is slung by the puborectalis component of the levator ani which pulls it forwards. The canal is approximately 4 cm long and angled postero-inferiorly. Developmentally, the midpoint of the anal canal is represented by the *dentate line*. This is the site at which the anal dimple in the skin, the ectodermal *proctodeum*, meets endoderm. This developmental implication is reflected by the following characteristics of the anal canal:

- The epithelium of the upper half of the anal canal is columnar. In contrast, the epithelium of the lower half of the anal canal is squamous. The mucosa of the upper canal is thrown into vertical columns (*of Morgagni*). At the bases of the columns are valve-like folds (*valves of Ball*). The level of the valves is termed the dentate line.
- The blood supply to the upper anal canal (see Fig. 16.5) is from the superior rectal artery (derived from the inferior mesenteric artery), whereas the lower anal canal is supplied by the inferior rectal artery (derived from the internal iliac artery). As mentioned previously, the venous drainage follows suit and represents a site of porto-systemic anastomosis (see p. 47).
- The upper anal canal is insensitive to pain as it is supplied by autonomic nerves only. The lower anal canal is sensitive to pain as it is supplied by somatic innervation (inferior rectal nerve).
- The lymphatics from the upper anal canal drain upwards along the superior rectal vessels to the internal iliac nodes, whereas lymph from the lower anal canal drains to the inguinal nodes.

The anal sphincter

See 'The anal region' Chapter 29 (p. 70).

Clinical notes

- **Appendicitis**: is the most commonly encountered surgical emergency in Western countries. It occurs mostly as a result of obstruction of the appendiceal lumen by a faecolith. Appendicectomy is performed most commonly through an oblique muscle-splitting incision in the right iliac fossa. The appendix is first located and then delivered into the wound. The mesentery of the appendix, containing the appendicular artery, is then ligated and divided. The appendix is tied at its base, excised and removed.
- **Colorectal cancer**: increasing age, the occidental low-fibre diet and genetic factors are predisposing factors to colorectal cancer. The majority occur in the recto-sigmoid region and, consequently, bleeding per rectum is a common presenting feature. Other presentations include anaemia, abdominal pain, bowel obstruction and perforation. Metastatic spread of the disease to the liver occurs as a result of tumour cell migration from the primary by way of the portal circulation. Systemic blood-borne metastases to the lungs and bones occur less frequently.
- **Diverticulosis**: occurs mostly in the sigmoid and descending colon. It is associated with advancing age and a low-fibre diet. In this condition, the colonic mucosa herniates through weaknesses in the circular muscle between the teniae coli. The points of maximal weakness tend to occur where blood vessels pierce the muscle. Diverticulosis is asymptomatic but, when symptoms arise as a result of inflammation, bleeding or perforation of a diverticulum, the condition is termed diverticular disease.

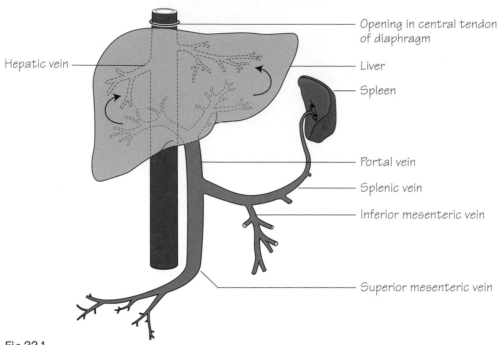

Opening in central tendon of diaphragm

Hepatic vein

Liver

Spleen

Portal vein

Splenic vein

Inferior mesenteric vein

Superior mesenteric vein

Fig.22.1
The venous circulation through the liver.
The transmission of blood from the portal system to the inferior vena cava
is via the liver lobules (fig. 22.2)

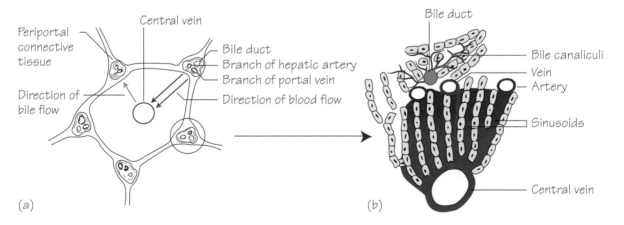

Central vein

Periportal connective tissue

Bile duct
Branch of hepatic artery
Branch of portal vein

Direction of bile flow

Direction of blood flow

Bile duct

Bile canaliculi

Vein

Artery

Sinusoids

Central vein

(a)

(b)

Fig.22.2
(a) A liver lobule to show the direction of blood flow from the portal system to the centrilobular veins
and thence to the inferior vena cava
(b) The blood flow through the sinusoids of the liver lobule and the passage of bile from the bile
canaliculi to the bile ducts

The liver (see Fig. 18.3)

• The liver predominantly occupies the right hypochondrium, but the left lobe extends to the epigastrium. Its domed upper (*diaphragmatic*) surface is related to the diaphragm and its lower border follows the contour of the right costal margin. When the liver is enlarged, the lower border becomes palpable below the costal margin.

• The liver anatomically consists of a large right lobe and a smaller left lobe. These are separated antero-superiorly by the falciform ligament and postero-inferiorly by *fissures for the ligamentum venosum and ligamentum teres*. Anatomically, the right lobe includes the caudate and quadrate lobes. The caudate lobe and the majority of the quadrate lobe are, however, functionally part of the left lobe, as they receive their blood supply from the left hepatic artery and deliver their bile into the left hepatic duct. The functional lobes of the liver are, therefore, separated by a vertical plane extending posteriorly from the gall bladder to the inferior vena cava (IVC).

• When the postero-inferior (*visceral*) surface of the liver is seen from behind, an H-shaped arrangement of grooves and fossae is identified. The boundaries of the H are formed as follows:

  • *Right anterior limb*: the gall-bladder fossa.
  • *Right posterior limb*: the groove for the IVC.
  • *Left anterior limb*: the fissure containing the ligamentum teres (the fetal remnant of the left umbilical vein which returns oxygenated blood from the placenta to the fetus).
  • *Left posterior limb*: the fissure for the ligamentum venosum (the latter structure is the fetal remnant of the ductus venosus; in the fetus, the ductus venosus serves to partially bypass the liver by transporting blood from the left umbilical vein to the IVC).
  • *Horizontal limb*: the *porta hepatis*. The caudate and quadrate lobes of the liver are the areas shown above and below the horizontal bar of the H, respectively.

• The *porta hepatis* is the hilum of the liver. It transmits (from posterior to anterior) the portal vein (Fig. 22.1), branches of the hepatic artery and hepatic ducts. The porta is enclosed within a double layer of peritoneum – the lesser omentum, which is firmly attached to the ligamentum venosum in its fissure.

• The liver is covered by peritoneum with the exception of the 'bare area'.

• The liver is made up of multiple functional units – *lobules* (Fig. 22.2). Branches of the portal vein and hepatic artery transport blood through portal canals into a central vein by way of sinusoids which traverse the lobules. The central veins ultimately coalesce into the right, left and central hepatic veins which drain blood from corresponding liver areas backwards into the IVC. The portal canals also contain tributaries of the hepatic ducts which serve to drain bile from the lobule down the biliary tree from where it can be concentrated in the gall bladder and eventually released into the duodenum. The extensive length of gut that is drained by the portal vein explains the predisposition for intestinal tumours to metastasise to the liver.

The gall bladder (see Fig. 18.3)

The gall bladder lies adherent to the undersurface of the liver in the transpyloric plane (p. 65) at the junction of the right and quadrate lobes. The duodenum and the transverse colon are behind it.

The gall bladder acts as a reservoir for bile which it concentrates. It usually contains approximately 50 mL of bile which is released through the cystic and then common bile ducts into the duodenum in response to gall-bladder contraction induced by gut hormones.

• **Structure**: the gall bladder comprises a *fundus*, a *body* and a *neck* (which opens into the cystic duct).

• **Blood supply**: the arterial supply to the gall bladder is derived from two sources: the cystic artery, which is usually, but not always, a branch of the right hepatic artery (Fig. 23.2), and small branches of the hepatic arteries, which pass via the fossa in which the gall bladder lies. The cystic artery represents the most significant source of arterial supply. There is, however, no corresponding cystic vein but venous drainage occurs via small veins passing through the gall-bladder bed.

The biliary tree (see Fig. 23.2)

The common hepatic duct is formed by the confluence of the right and left hepatic ducts in the porta hepatis. The common hepatic duct is joined by the cystic duct to form the common bile duct. This structure courses, sequentially, in the free edge of the lesser omentum, behind the first part of the duodenum and in the groove between the second part of the duodenum and the head of the pancreas. It ultimately opens at the papilla on the medial aspect of the second part of the duodenum.

The common bile duct usually, but not always, joins with the main *pancreatic duct (of Wirsung)* (p. 59).

Clinical notes

• **Cholelithiasis**: gallstones are composed of cholesterol, bile pigment or, more commonly, a mixture of these two constituents. Cholesterol stones form due to an altered composition of bile resulting in the precipitation of cholesterol crystals. Most gallstones are asymptomatic; however, when they block the exit from the gall bladder, or migrate down the biliary tree, they can be responsible for a diverse array of complications, such as acute cholecystitis, biliary colic, cholangitis and pancreatitis.

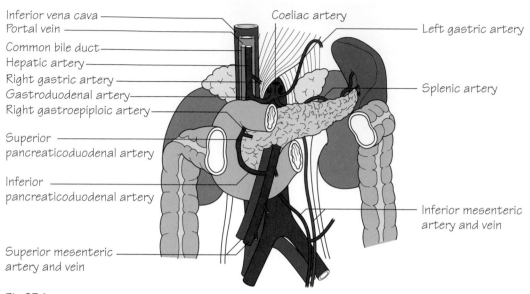

Inferior vena cava
Portal vein
Common bile duct
Hepatic artery
Right gastric artery
Gastroduodenal artery
Right gastroepiploic artery
Superior pancreaticoduodenal artery
Inferior pancreaticoduodenal artery
Superior mesenteric artery and vein

Coeliac artery
Left gastric artery
Splenic artery
Inferior mesenteric artery and vein

Fig.23.1
The relations of the pancreas

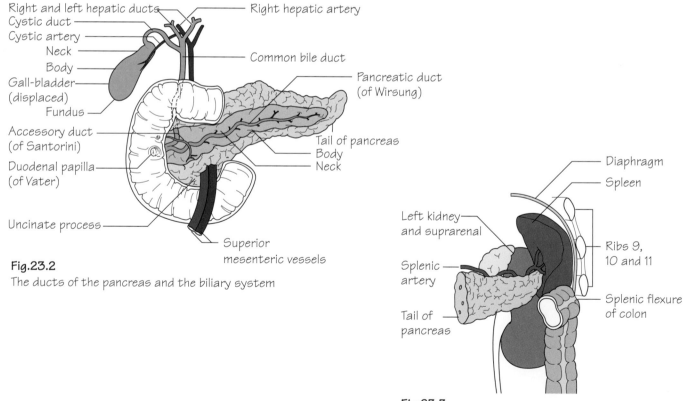

Right and left hepatic ducts
Cystic duct
Cystic artery
Neck
Body
Gall-bladder (displaced)
Fundus
Accessory duct (of Santorini)
Duodenal papilla (of Vater)
Uncinate process

Right hepatic artery
Common bile duct
Pancreatic duct (of Wirsung)
Tail of pancreas
Body
Neck
Superior mesenteric vessels

Fig.23.2
The ducts of the pancreas and the biliary system

Left kidney and suprarenal
Splenic artery
Tail of pancreas

Diaphragm
Spleen
Ribs 9, 10 and 11
Splenic flexure of colon

Fig.23.3
The relations of the spleen

The pancreas (Figs. 23.1 and 23.2)

The pancreas has a *head*, *neck*, *body* and *tail*. It is a retroperitoneal organ that lies roughly along the transpyloric plane. The head is bound laterally by the curved duodenum and the tail extends to the hilum of the spleen in the lienorenal ligament. The superior mesenteric vessels pass behind the pancreas, then anteriorly over the *uncinate process* and third part of the duodenum into the root of the small bowel mesentery. The inferior vena cava, aorta, coeliac plexus, left kidney (and its vessels) and left adrenal gland are posterior pancreatic relations. In addition, the portal vein is formed behind the pancreatic neck by the confluence of the splenic and superior mesenteric veins. The lesser sac and stomach are anterior pancreatic relations.

- **Structure**: the main *pancreatic duct (of Wirsung)* courses the length of the gland, ultimately draining pancreatic secretions into the *ampulla of Vater*, together with the common bile duct, and thence into the second part of the duodenum. An *accessory duct (of Santorini)* drains the uncinate process of the pancreas, opening slightly proximal to the ampulla into the second part of the duodenum.
- **Blood supply**: the pancreatic head receives its supply from the superior and inferior pancreaticoduodenal arteries. The splenic artery courses along the upper border of the body of the pancreas which it supplies by means of a large branch – the *arteria pancreatica magna* – and numerous smaller branches.
- **Function**: the pancreas is a lobulated structure which performs both exocrine and endocrine functions. The exocrine secretory glands drain pancreatic juice into the pancreatic ducts and, from there, ultimately into the duodenum. The secretion is essential for the digestion and absorption of proteins, fats and carbohydrates. The endocrine pancreas is responsible for the production and secretion of glucagon and insulin, which take place in specialised cells of the islets of Langerhans.

The spleen (Fig. 23.3)

The spleen is approximately the size of a clenched fist and lies directly below the left hemidiaphragm which, in addition to the pleura, separates it from the overlying 9th, 10th and 11th ribs.

- **Peritoneal attachments**: the splenic capsule is fibrous with peritoneum adherent to its surface. The gastrosplenic and lienorenal ligaments attach it to the stomach and kidney, respectively. The former ligament carries the short gastric and left gastroepiploic vessels to the fundus and greater curvature of the stomach, and the latter ligament carries the splenic vessels and tail of the pancreas towards the left kidney.
- **Blood supply**: is from the splenic artery to the hilum of the spleen. Venous drainage is to the splenic vein, thence to the portal vein.
- **Structure**: the spleen is a highly vascular reticulo-endothelial organ. It consists of a thin capsule from which trabeculae extend into the splenic pulp. In the spleen, the immunological centres, i.e. the lymphoid follicles (the *white pulp*), are scattered throughout richly vascularised sinusoids (the *red pulp*).

Clinical notes

- **Acute pancreatitis**: the presence of gallstones or a history of excessive alcohol intake are most commonly associated with pancreatitis. The mechanism by which these aetiological factors result in pancreatic injury is unknown; however, they both appear to lead to the activation of pancreatic exocrine pro-enzymes with resultant autodigestion. Even today, the mortality rate for severe acute pancreatitis is 10–20%.
- **Splenectomy**: as the spleen is a highly vascular organ, any injury to it can be life-threatening. Under these circumstances, splenectomy must be carried out urgently. The technique used differs slightly when the procedure is performed for emergency as opposed to elective indications, but the principles are similar. Splenectomy involves ligature of the splenic vessels approaching the hilum (taking care not to injure the tail of the pancreas or colon) and dissection of the splenic pedicles – the gastro-splenic (including the short gastric vessels) and lienorenal ligaments. As the spleen is an important immunological organ, patients are rendered immunocompromised following splenectomy, and are particularly vulnerable to infection with capsulated bacteria (e.g. meningococcus, pneumococcus) as these organisms are principally eliminated at splenic lymphoid follicles. Patients about to undergo splenectomy are, therefore, routinely vaccinated against the capsulated bacteria and children, who are at the greatest risk of sepsis, are maintained on long-term antibiotic prophylaxis.

24 The posterior abdominal wall

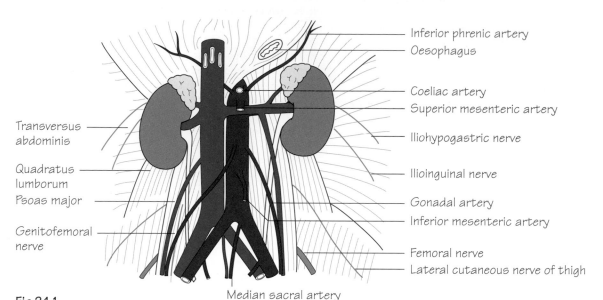

Inferior phrenic artery
Oesophagus
Coeliac artery
Superior mesenteric artery
Iliohypogastric nerve
Ilioinguinal nerve
Gonadal artery
Inferior mesenteric artery
Femoral nerve
Lateral cutaneous nerve of thigh

Transversus abdominis
Quadratus lumborum
Psoas major
Genitofemoral nerve

Median sacral artery

Fig.24.1
The structures of the posterior abdominal wall

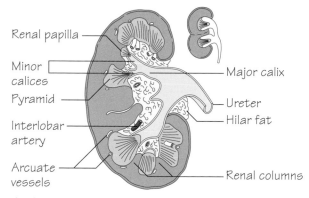

Renal papilla
Minor calices
Pyramid
Interlobar artery
Arcuate vessels

Major calix
Ureter
Hilar fat
Renal columns

Fig.24.2
A section through the right kidney.
The small diagram shows how the renal columns represent the cortices of adjacent fused lobes

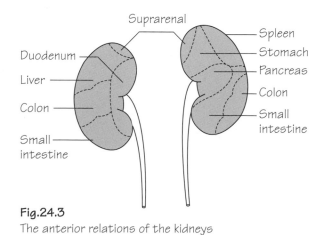

Suprarenal
Duodenum
Liver
Colon
Small intestine

Spleen
Stomach
Pancreas
Colon
Small intestine

Fig.24.3
The anterior relations of the kidneys

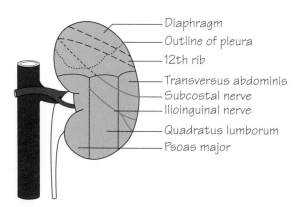

Diaphragm
Outline of pleura
12th rib
Transversus abdominis
Subcostal nerve
Ilioinguinal nerve
Quadratus lumborum
Psoas major

Fig.24.4
The posterior relations of the kidneys

Anatomy at a Glance, Third Edition. Omar Faiz, Simon Blackburn and David Moffat.
© 2011 Blackwell Publishing Ltd. Published 2011 by Blackwell Publishing Ltd.

The structures of the posterior abdominal wall (Fig. 24.1)

These include:

- *Muscles*, including psoas major and quadratus lumborum (see Muscle index, p. 178).
- The *abdominal aorta* and its branches (see p. 43).
- The *inferior vena cava* (IVC) and its tributaries (see p. 47).
- The *kidneys*.
- The *ureters*.
- The *adrenal (suprarenal) glands*.
- The *lumbar sympathetic trunks and plexuses* and the *lumbar plexus* (see p. 63).

The kidneys (Fig. 24.2)

- **Structure**: the kidney has its own fibrous capsule and is surrounded by *perinephric fat* which, in turn, is enclosed by *renal fascia*. Each kidney is approximately 10–12 cm long and consists of an outer *cortex*, an inner *medulla* and a *pelvis*.

The hilum of the kidney is situated medially and transmits, from front to back, the renal vein, renal artery, ureteric pelvis as well as lymphatics and sympathetic vasomotor nerves.

The renal pelvis divides into two or three major calices and these, in turn, divide into minor calices which receive urine from the medullary pyramids by way of the papillae.

- **Position**: the kidneys lie in the retroperitoneum against the posterior abdominal wall. The right kidney lies approximately 1 cm lower than the left.
- **Relations**: see Figs. 24.3 and 24.4.
- **Blood supply**: the renal arteries arise from the aorta at the level of L2. Together, the renal arteries direct 25% of the cardiac output towards the kidneys. Each renal artery divides into five *segmental arteries* at the hilum which, in turn, divide sequentially into *lobar*, *interlobar*, *arcuate* and *cortical radial branches*. The cortical radial branches give rise to the afferent arterioles which supply the glomeruli and extend further to become efferent arterioles. The differential pressures between afferent and efferent arterioles lead to the production of an ultrafiltrate which then passes through, and is modified by, the nephron to produce urine.

The right renal **artery** passes behind the IVC. The left renal **vein** is long as it courses in front of the aorta to drain into the IVC.

- **Lymphatic drainage**: to the para-aortic lymph nodes.

The ureter (Fig. 24.1)

The ureter is considered in abdominal, pelvic and intravesical portions.

- **Structure**: in the adult male, the ureter is approximately 20–30 cm long and courses from the hilum of the kidney to the bladder. It has a muscular wall and is lined by transitional epithelium. At operation, it can be recognized by its worm–like peristalsis, which is termed *vermiculation*.

- **Course**: from the renal pelvis at the hilum, the course of the ureter can be summarised as follows:
  - It passes along the medial part of psoas major behind, but adherent to, the peritoneum.
  - It, then, crosses the common iliac bifurcation anterior to the sacro-iliac joint and courses over the lateral wall of the pelvis to the ischial spine.
  - At the ischial spine, the ureter passes forwards and medially to enter the bladder obliquely. The intravesical portion of the ureter is approximately 2 cm long and its passage through the bladder wall produces a sphincter-like effect. In the male, the ureter is crossed superficially near its termination by the *vas deferens*. In the female, the ureter passes above the lateral fornix of the vagina but below the broad ligament and uterine vessels.
- **Blood supply**: as the ureter is an abdominal and pelvic structure, it receives a blood supply from multiple sources:
  - The *upper ureter*: receives direct branches from the aorta, renal and gonadal arteries.
  - The *lower ureter*: receives branches of the internal iliac and inferior vesical arteries.

Adrenal (suprarenal) glands (Fig. 24.3)

The adrenal glands comprise an outer *cortex* and inner *medulla*. The cortex is derived from mesoderm and is responsible for the production of steroid hormones (glucocorticoids, mineralocorticoids and sex steroids). The medulla is derived from ectoderm (neural crest) and acts as a part of the autonomic nervous system. It receives sympathetic preganglionic fibres from the greater splanchnic nerves which stimulate the medulla to secrete noradrenaline and adrenaline into the bloodstream.

- **Position**: the adrenals are small glands which lie in the renal fascia on the upper poles of the kidneys. The right gland lies behind the right lobe of the liver and immediately posterolateral to the IVC. The left adrenal is anteriorly related to the lesser sac and stomach.
- **Blood supply**: the phrenic, renal arteries and aorta all contribute branches to the adrenal glands. Venous drainage is on the right to the IVC and on the left to the left renal vein.

Clinical notes

- **Ureteric stones**: most ureteric calculi arise for unknown reasons, although inadequate urinary drainage, the presence of infected urine and hypercalcaemia are definite predisposing factors. The presence of an impacted ureteric stone is characterised by haematuria and agonising colicky pain (ureteric colic), which classically radiates from loin to groin. Large impacted stones can lead to hydronephrosis and/or infection of the affected kidney and, consequently, need to be broken down or removed by interventional or open procedures.

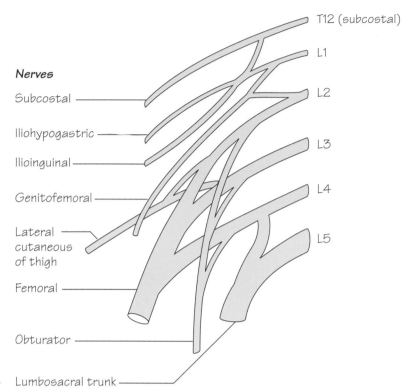

Nerves

Subcostal

Iliohypogastric

Ilioinguinal

Genitofemoral

Lateral cutaneous of thigh

Femoral

Obturator

T12 (subcostal)

L1

L2

L3

L4

L5

Fig.25.1
The lumbar plexus — Lumbosacral trunk

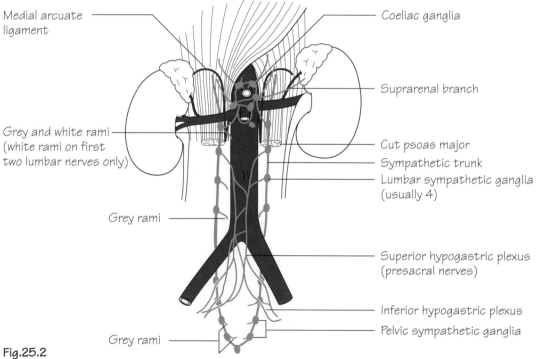

Medial arcuate ligament

Grey and white rami (white rami on first two lumbar nerves only)

Grey rami

Grey rami

Coeliac ganglia

Suprarenal branch

Cut psoas major

Sympathetic trunk

Lumbar sympathetic ganglia (usually 4)

Superior hypogastric plexus (presacral nerves)

Inferior hypogastric plexus

Pelvic sympathetic ganglia

Fig.25.2
The sympathetic system in the abdomen and pelvis

Anatomy at a Glance, Third Edition. Omar Faiz, Simon Blackburn and David Moffat.

The lumbar plexus (Fig. 25.1)

- The lumbar plexus is formed from the anterior primary rami of L1–4. The trunks of the plexus lie within the substance of psoas major and, with the exceptions of the obturator and genitofemoral nerves, emerge at its lateral border.
- The 12th intercostal nerve is also termed the *subcostal nerve* as it has no intercostal space but, instead, runs below the rib in the neurovascular plane to supply the abdominal wall.
- The *iliohypogastric nerve* is the main trunk of the 1st lumbar nerve. It supplies the skin of the upper buttock by way of a lateral cutaneous branch, and terminates by piercing the external oblique above the superficial inguinal ring, where it supplies the overlying skin of the mons pubis. The *ilioinguinal nerve* is the collateral branch of the iliohypogastric. The ilioinguinal nerve runs in the neurovascular plane of the abdominal wall to emerge through the superficial inguinal ring to provide a cutaneous supply to the skin of the medial thigh, the root of the penis and the anterior one-third of the scrotum (or labium majus in the female).
- The *genitofemoral nerve* (L1, 2) emerges from the anterior surface of psoas major. It courses inferiorly and divides into a genital component that enters the spermatic cord and supplies the cremaster (in the male) and a femoral component that supplies the skin of the thigh overlying the femoral triangle.
- The *lateral cutaneous nerve of the thigh* (L2, 3), having emerged from the lateral border of psoas major, encircles the iliac fossa to pass under the inguinal ligament (p. 113).
- The *femoral nerve* (L2–4, posterior division): see p. 113.
- The *obturator nerve* (L2–4, anterior division): see p. 113.
- A large part of L4 joins with L5 to contribute to the sacral plexus as the lumbosacral trunk.

Lumbar sympathetic chain (Fig. 25.2)

- **Sympathetic supply**: the lumbar sympathetic chain is a continuation of the thoracic sympathetic chain as it passes under the medial arcuate ligament of the diaphragm. The chain passes anterior to the lumbar vertebral bodies and usually carries four ganglia, which send grey rami communicantes to the lumbar spinal nerves. The upper two ganglia receive white rami from L1 and L2.

The lumbar sympathetic chain, the splanchnic nerves and the vagus contribute sympathetic and parasympathetic branches to plexuses (coeliac, superior mesenteric, renal and inferior mesenteric) around the abdominal aorta. In addition, other branches continue inferiorly to form the *superior hypogastric plexus* (*presacral nerves*) from where they branch into right and left *inferior hypogastric plexuses*. The latter also receive a parasympathetic supply from the pelvic splanchnic nerves. The branches from the inferior hypogastric plexuses are distributed to the pelvic viscera along the course and branches of the internal iliac artery.

The *coeliac ganglia* are prominent and lie around the origins of the coeliac and superior mesenteric arteries.

- **Parasympathetic supply**: the pelvic viscera arises from the anterior primary rami of S2,3,4–the *pelvic splanchnic nerves*. The latter parasympathetic supply reaches proximally as far as the junction between the hindgut and midgut on the transverse colon.

Clinical notes

- **Lumbar sympathectomy**: this procedure is performed in cases of severe peripheral vascular disease of the lower limbs where vascular reconstructive surgery is not possible and skin necrosis is imminent. The procedure aims to improve the cutaneous blood supply. It is ineffective in treating lower limb claudication. Sympathectomy can be performed surgically or chemically. The operation involves excision of the 2nd–4th lumbar ganglia with the intermediate chain. Chemical sympathectomy involves injection of phenol into the region of the lumbar sympathetic chain under radiological guidance.

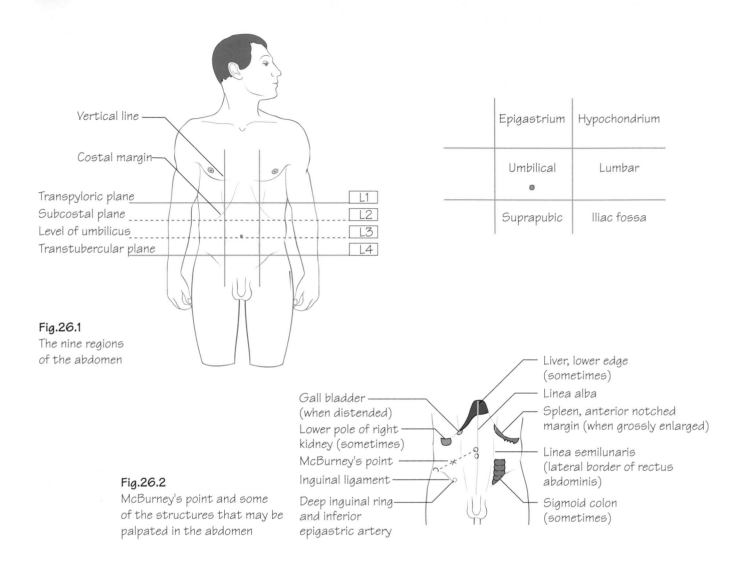

Epigastrium	Hypochondrium
Umbilical	Lumbar
Suprapubic	Iliac fossa

Fig.26.1
The nine regions
of the abdomen

Fig.26.2
McBurney's point and some
of the structures that may be
palpated in the abdomen

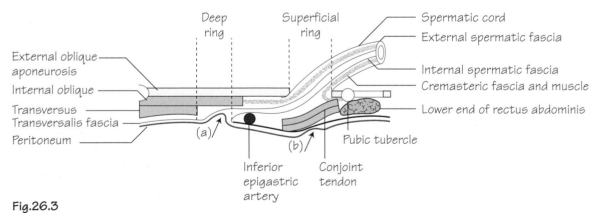

Fig.26.3
A horizontal section through the inguinal canal. Diagrammatic.
(a) and (b) show the sites of indirect and direct herniae respectively

Vertebral levels (Fig. 26.1)

(In each case the lower border is referred to.)
- **T9**: *xiphoid process.*
- **L1**: the *transpyloric plane (of Addison).* This horizontal plane lies halfway between the suprasternal notch and the symphysis pubis, and passes approximately through the tip of the 9th costal cartilage, the pylorus, pancreatic neck, duodenojejunal flexure, the gall-bladder fundus and the hila of the kidneys. This plane also corresponds to the level at which the spinal cord terminates and the point at which the lateral edge of the rectus abdominis crosses the costal margin.
- **L2**: *the subcostal plane.* This plane corresponds to the line joining the lowest points of the thoracic cage—the lower margin of the 10th rib laterally.
- **L3**: the level of the umbilicus (in a young slim person).
- **L4**: the *transtubercular plane.* This corresponds to the line which joins the tubercles of the iliac crests.

Lines of orientation

Vertical lines: these are imaginary and are most often used with the subcostal and intertubercular planes, for purposes of description, to subdivide the abdomen into nine regions (Fig. 26.1). They pass vertically, on either side, through the point halfway between the anterior superior iliac spine and the pubic tubercle. More commonly used, for description of pain location, are quadrants. The latter are imaginary lines arising by the bisection of the umbilicus by vertical and horizontal lines.

Surface markings of the abdominal wall

- The *costal margin* (Fig. 26.1) is the inferior margin of the thoracic cage. It includes the costal cartilages anteriorly, the 7th–10th ribs laterally and the cartilages of the 11th and 12th ribs posteriorly.
- The *symphysis pubis* is an easily palpable secondary cartilaginous joint which lies between the pubic bones in the midline. The *pubic tubercle* is an important landmark and is identifiable on the superior surface of the pubis.
- The *inguinal ligament* (Figs. 15.1 and 26.2) is attached laterally to the anterior superior iliac spine and medially to the pubic tubercle.
- The *superficial inguinal ring* (see Fig. 15.1) is a triangular-shaped defect in the external oblique aponeurosis. It is situated above and medial to the pubic tubercle.
- The *spermatic cord* can be felt passing medially to the pubic tubercle and descending into the scrotum.
- The *deep inguinal ring* (Fig. 26.3) lies halfway along the line from the anterior superior iliac spine to the pubic tubercle.
- The *linea alba* (see Fig. 15.1) is formed by the fusion of the aponeuroses of the muscles of the anterior abdominal wall. It extends as a depression in the midline from the xiphoid process to the symphysis pubis.
- The *linea semilunaris* is the lateral edge of the rectus abdominis muscle. It crosses the costal margin at the tip of the 9th costal cartilage.

Surface markings of the abdominal viscera (Fig. 26.2)

- **Liver**: the lower border of the liver is usually just palpable on deep inspiration in slim individuals. The upper border follows the undersurface of the diaphragm and reaches a level just below the nipple on each side.
- **Spleen**: this organ lies below the left hemidiaphragm deep to the 9th, 10th and 11th ribs posteriorly. The *anterior notch* reaches the mid-axillary line anteriorly.
- **Gall bladder**: the fundus of the gall bladder lies in the transpyloric plane (L1). The surface marking corresponds to a point at which the lateral border of rectus abdominis (*linea semilunaris*) crosses the costal margin.
- **Pancreas**: the pancreatic neck lies on the level of the transpyloric plane (L1). The pancreatic head lies to the right and below the neck, whereas the body and tail pass upwards and to the left.
- **Aorta**: the aorta bifurcates to the left of the midline at the level of L4.
- **Kidneys**: the kidney hila lie on the level of the transpyloric plane (L1). The lower pole of the right kidney usually extends 3 cm below the level of the left and is often palpable in slim subjects.
- **Appendix**: *McBurney's point* represents the surface marking for the base of the appendix. This point lies one-third of the way along the line joining the anterior superior iliac spine and the umbilicus. McBurney's point is important surgically as it represents the usual site of maximal tenderness in appendicitis and also serves as the central point for the incision made when performing an appendicectomy.
- **Bladder**: in adults, the bladder is a pelvic organ and can be palpated above the symphysis pubis only when full or enlarged.

Clinical notes

- **Midline laparotomy incision**: this incision is the most commonly used incision for accessing the intraperitoneal contents. The layers incised in this approach include the skin, linea alba, transversalis fascia and extraperitoneal fat, before the peritoneum itself is reached.
- **Inguinal herniae** (Figs. 26.3 and 57.1):
  - *Indirect inguinal herniae*: arise in childhood as a result of persistence of the processus vaginalis of the fetus. Indirect herniae can, however, occur at any age. Abdominal contents bulge through the deep inguinal ring into the canal and, eventually, into the scrotum. This type of hernia can be controlled by digital pressure over the deep ring.
  - *Direct inguinal herniae*: arise as a result of weakness in the posterior wall of the inguinal canal. This type of hernia cannot be controlled by digital pressure over the deep ring and rarely does pass into the scrotum.

  The clinical distinction between direct and indirect inguinal herniae can be difficult. At operation, however, the relation of the hernial neck to the inferior epigastric artery defines the hernia type, i.e. the neck of the sac of an indirect hernia lies lateral to the artery, whereas that of a direct type lies medial to it.

27 The pelvis I – the bony and ligamentous pelvis

Iliac crest
Inferior gluteal line
Anterior gluteal line
Anterior superior iliac spine
Posterior superior iliac spine
Anterior inferior iliac spine
Posterior gluteal line
Acetabulum
Greater sciatic notch
Obturator foramen
Pubic tubercle
Spine of ischium
Pubic crest
Lesser sciatic notch
Body of pubis
Ischial tuberosity
Inferior ramus
Ramus of ischium

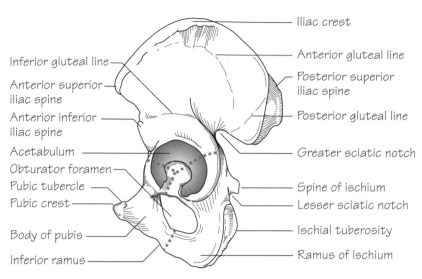

Fig.27.1
The lateral surface of the left hip bone

Iliac fossa
Auricular surface
Iliopectineal line
Pubic tubercle
Pubic symphysis

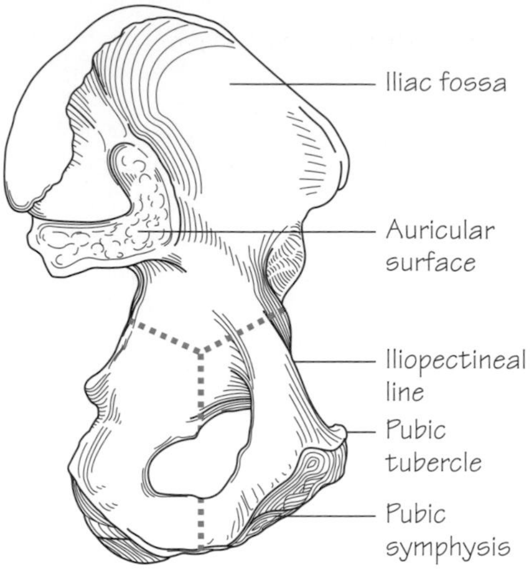

Fig.27.2
The medial surface of the left hip bone

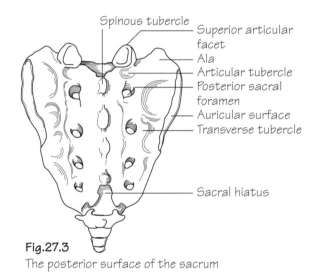

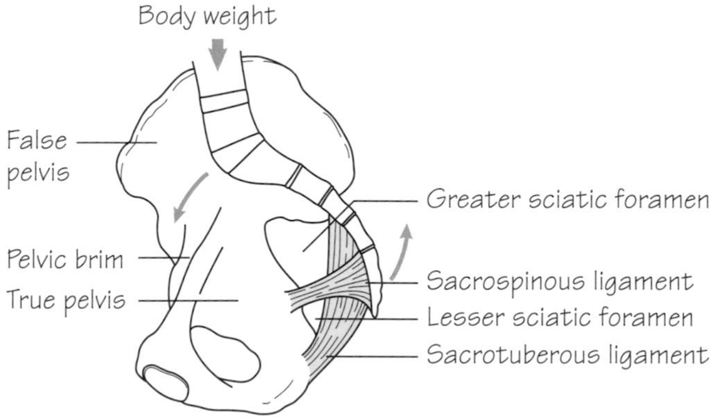

Spinous tubercle
Superior articular facet
Ala
Articular tubercle
Posterior sacral foramen
Auricular surface
Transverse tubercle
Sacral hiatus

Fig.27.3
The posterior surface of the sacrum

Body weight
False pelvis
Greater sciatic foramen
Pelvic brim
Sacrospinous ligament
True pelvis
Lesser sciatic foramen
Sacrotuberous ligament

Fig.27.4
The sacrospinous and sacrotuberous ligaments resist rotation of the sacrum due to the body weight

Fig.27.5
The male pelvic floor from above. The blue line represents the origin of levator ani from the obturator fascia

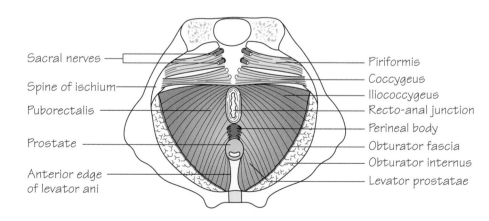

Sacral nerves
Piriformis
Coccygeus
Spine of ischium
Iliococcygeus
Puborectalis
Recto-anal junction
Perineal body
Prostate
Obturator fascia
Obturator internus
Anterior edge of levator ani
Levator prostatae

The pelvis is bounded posteriorly by the sacrum and coccyx and anterolaterally by the innominate bones.

Os innominatum (hip bone) (Figs. 27.1 and 27.2)

This bone comprises three component parts: the *ilium*, *ischium* and *pubis*. By adulthood, the constituent bones have fused together at the acetabulum. Posteriorly, each hip bone articulates with the sacrum at the *sacro-iliac joint* (a synovial joint).

• **Ilium**: the *iliac crest* forms the upper border of the bone. It runs backwards from the *anterior superior iliac spine* to the *posterior superior iliac spine*. Below each of these bony landmarks are the corresponding inferior spines. The outer surface of the ilium is termed the *gluteal surface* as it is where the gluteal muscles are attached. The *inferior*, *anterior* and *posterior gluteal lines* demarcate the bony attachments of the glutei. The inner surface of the ilium is smooth and hollowed out to form the *iliac fossa*. It gives attachment to the iliacus muscle. The *auricular* surface of the ilium articulates with the sacrum at the *sacro-iliac joints* (synovial joints). *Posterior*, *interosseous* and *anterior sacro-iliac ligaments* strengthen the sacro-iliac joints. The *iliopectineal line* courses anteriorly on the inner surface of the ilium from the auricular surface to the pubis. It forms the lateral margin of the *pelvic brim* (see below).

• **Ischium**: comprises a *spine* on its posterior part which demarcates the *greater* (above) and *lesser* (below) *sciatic notches*. The *ischial tuberosity* is a thickening on the lower part of the body of the ischium which bears weight in the sitting position. The *ischial ramus* projects forwards from the tuberosity to meet and fuse with the *inferior pubic ramus*.

• **Pubis**: comprises a *body* and *superior* and *inferior pubic rami*. It articulates with the pubic bone of the other side at the *symphysis pubis* (a secondary cartilaginous joint). The superior surface of the body bears the *pubic crest* and the *pubic tubercle* (Fig. 27.1).

The *obturator foramen* is a large opening bounded by the rami of the pubis and ischium.

The sacrum and coccyx (Fig. 27.3)

• The sacrum comprises five fused vertebrae. The anterior and lateral aspects of the sacrum are termed the *central* and *lateral masses*, respectively. The upper anterior part is termed the *sacral promontory*. Four *anterior sacral foramina* on each side transmit the upper four sacral anterior primary rami. Posteriorly, the fused pedicles and laminae form the *sacral canal*, representing a continuation of the vertebral canal. Inferiorly, the canal terminates at the *sacral hiatus*. *Sacral cornua* bound the hiatus inferiorly on either side. The subarachnoid space terminates at the level of S2. The sacrum is tilted anteriorly to form the *lumbosacral angle* with the lumbar vertebra.

• The coccyx articulates superiorly with the sacrum. It comprises between three and five fused rudimentary vertebrae.

The obturator membrane

The obturator membrane is a sheet of fibrous tissue which covers the obturator foramen, with the exception of a small area for the passage of the obturator nerve and vessels which traverse the canal to pass from the pelvis to gain access to the thigh.

The pelvic cavity

The *pelvic brim* (also termed the pelvic inlet) separates the pelvis into the *false pelvis* (above) and the *true pelvis* (below). The brim is formed by the sacral promontory behind, the iliopectineal lines laterally and the symphysis pubis anteriorly. The *pelvic outlet* is bounded by the coccyx behind, the ischial tuberosities laterally and the *pubic arch* anteriorly. The true pelvis (pelvic cavity) lies between the inlet and outlet. The false pelvis is best considered as part of the abdominal cavity.

The ligaments of the pelvis (Fig. 27.4)

These include the:

• **Sacrotuberous ligament**: extends from the lateral part of the sacrum and coccyx to the ischial tuberosity.

• **Sacrospinous ligament**: extends from the lateral part of the sacrum and coccyx to the ischial spine.

The above ligaments, together with the sacro-iliac ligaments, bind the sacrum and coccyx to the os and prevent excessive movement at the sacro-iliac joints. In addition, these ligaments create the *greater* and *lesser sciatic foramina* from the greater and lesser sciatic notches.

The pelvic floor (Fig. 27.5)

The pelvic floor muscles support the viscera, produce a sphincter action on the rectum and vagina, and help to produce increases in intra-abdominal pressure during straining. The rectum, urethra (and vagina in the female) traverse the pelvic floor to gain access to the exterior. The levator ani and coccygeus muscles form the pelvic floor, while piriformis covers the front of the sacrum.

• **Levator ani**: arises from the posterior aspect of the pubis, the fascia overlying the obturator internus on the side wall of the pelvis and the ischial spine. From this broad origin fibres sweep backwards towards the midline as follows:

• *Anterior fibres (sphincter vaginae or levator prostatae)*: these fibres surround the vagina in the female (prostate in the male) and insert into the *perineal body*. The latter structure is a fibromuscular node which lies anterior to the anal canal.

• *Intermediate fibres (puborectalis)*: these fibres surround the anorectal junction and also insert into the deep part of the anal sphincter. They provide an important voluntary sphincter action at the anorectal junction.

• *Posterior fibres (iliococcygeus)*: these fibres insert into the lateral aspect of the coccyx and a median fibrous raphe (the anococcygeal body).

• **Coccygeus**: arises from the ischial spine and inserts into the lower sacrum and coccyx.

Sex differences in the pelvis

The female pelvis differs from that of the male for the purpose of childbearing. The major sex differences include:

1 The pelvic inlet is oval in the female. In the male, the sacral promontory is more prominent, producing a heart-shaped inlet.

2 The pelvic outlet is wider in females as the ischial tuberosities are everted.

3 The pelvic cavity is more spacious in the female than in the male.

4 The false pelvis is shallow in the female.

5 The pubic arch (the angle between the inferior pubic rami) is wider and more rounded in the female when compared with that of the male.

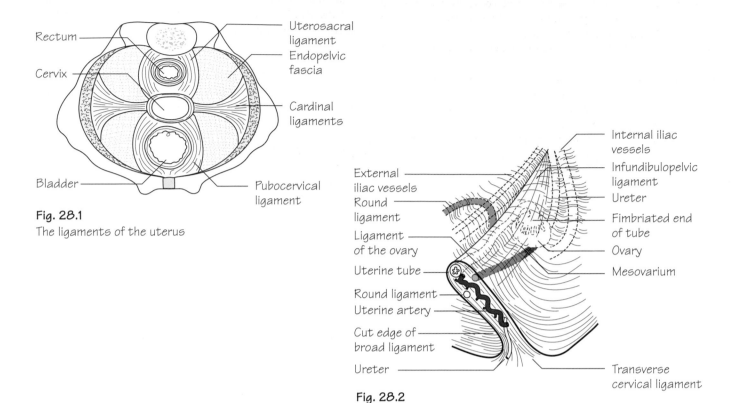

Rectum

Cervix

Bladder

Uterosacral ligament

Endopelvic fascia

Cardinal ligaments

Pubocervical ligament

Fig. 28.1
The ligaments of the uterus

External iliac vessels

Round ligament

Ligament of the ovary

Uterine tube

Round ligament

Uterine artery

Cut edge of broad ligament

Ureter

Internal iliac vessels

Infundibulopelvic ligament

Ureter

Fimbriated end of tube

Ovary

Mesovarium

Transverse cervical ligament

Fig. 28.2
The broad ligament cut off close to the uterus

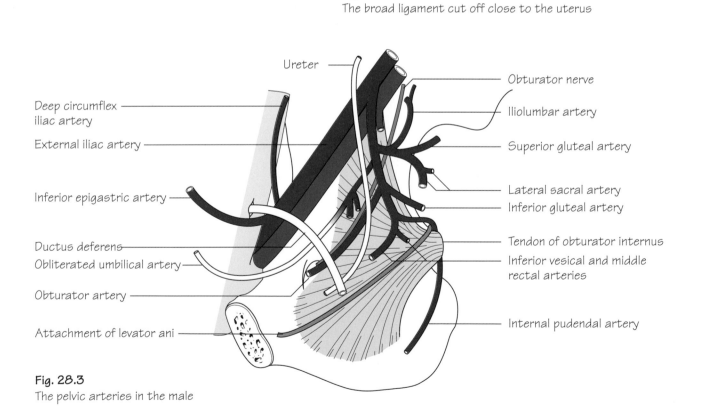

Ureter

Deep circumflex iliac artery

External iliac artery

Inferior epigastric artery

Ductus deferens

Obliterated umbilical artery

Obturator artery

Attachment of levator ani

Obturator nerve

Iliolumbar artery

Superior gluteal artery

Lateral sacral artery

Inferior gluteal artery

Tendon of obturator internus

Inferior vesical and middle rectal arteries

Internal pudendal artery

Fig. 28.3
The pelvic arteries in the male

Anatomy at a Glance, Third Edition. Omar Faiz, Simon Blackburn and David Moffat.
© 2011 Blackwell Publishing Ltd. Published 2011 by Blackwell Publishing Ltd.

Pelvic fascia (Fig. 28.1)

The *pelvic fascia* is the term given to the connective tissue that lines the pelvis covering levator ani and obturator internus. It is continuous with the fascial layers of the abdominal wall above and the perineum below. *Endopelvic fascia* is the term given to the loose connective tissue that covers the pelvic viscera. The endopelvic fascia is condensed into fascial ligaments which act as supports for the cervix and vagina. These ligaments include the:

- **Cardinal (Mackenrodt's) ligaments**: pass laterally from the cervix and upper vagina to the pelvic side walls.
- **Uterosacral ligaments**: pass backwards from the cervix and vaginal fornices to the fascia overlying the sacro-iliac joints.
- **Pubocervical ligaments**: extend anteriorly from the cardinal ligaments to the pubis (puboprostatic in the male).
- **Pubovesical ligaments**: run from the back of the symphysis pubis to the bladder neck.

The broad and round ligaments (Fig. 28.2)

- **Broad ligament**: is a double fold of peritoneum which hangs between the lateral aspect of the uterus and the pelvic side walls. The ureter passes forwards under this ligament, but above and lateral to the lateral fornix of the vagina, to gain access to the bladder. The broad ligament contains the following structures:
  - Fallopian tube.
  - Ovary.
  - Ovarian ligament.
  - Round ligament (see below).
  - Uterine and ovarian vessels.
  - Nerves and lymphatics.
- **Round ligament**: is a cord-like fibromuscular structure which is the female equivalent of the gubernaculum in the male. It passes from the lateral angle of the uterus to the labium majus by coursing in the broad ligament and then through the inguinal canal (p. 42).

Arteries of the pelvis (Fig. 28.3)

- **Common iliac arteries**: arise from the aortic bifurcation to the left of the midline at the level of the umbilicus. These arteries, in turn, bifurcate into external and internal iliac branches anterior to the sacro-iliac joints on either side.
- **External iliac artery**: courses from its origin (described above) to become the femoral artery as it passes under the inguinal ligament at the mid-inguinal point. The external iliac artery gives rise to branches which supply the anterior abdominal wall. These include the *deep circumflex iliac artery* and *inferior epigastric artery*. The latter branch gains access to the rectus sheath, which it supplies, and eventually anastomoses with the superior epigastric artery.
- **Internal iliac artery**: courses from its origin (described above) to divide into anterior and posterior trunks at the level of the greater sciatic foramen.

Branches of the anterior trunk

- **Umbilical artery**: although the distal part is obliterated, the proximal part is patent and gives rise to the *superior vesical artery*, which contributes a supply to the bladder.
- **Obturator artery**: passes with the obturator nerve through the obturator canal to enter the thigh.
- **Inferior vesical artery**: as well as contributing a supply to the bladder, it also gives off a *branch to the vas deferens* (in the male).
- **Middle rectal artery**: anastomoses with the superior and inferior rectal arteries to supply the rectum.
- **Internal pudendal artery**: is the predominant supply to the perineum. It exits the pelvis briefly through the greater sciatic foramen, but then re-enters below piriformis through the lesser sciatic foramen to enter the pudendal canal together with the pudendal nerve.
- **Uterine artery**: passes medially on the pelvic floor and then over the ureter and lateral fornix of the vagina to ascend the lateral aspect of the uterus between the layers of the broad ligament.
- **Inferior gluteal artery**: passes out of the pelvis through the greater sciatic foramen to the gluteal region which it supplies.
- **Vaginal artery**.

Branches of the posterior trunk

- **Superior gluteal artery**: contributes a supply to the gluteal muscles. It leaves the pelvis through the greater sciatic foramen.
- **Iliolumbar artery**.
- **Lateral sacral artery**.

Veins of the pelvis

The right and left common iliac veins join to form the inferior vena cava behind the right common iliac artery but anterolateral to the body of L5. The overall arrangement of pelvic venous drainage reciprocates that of the arterial supply.

Nerves of the pelvis

Sacral plexus (see p. 115).

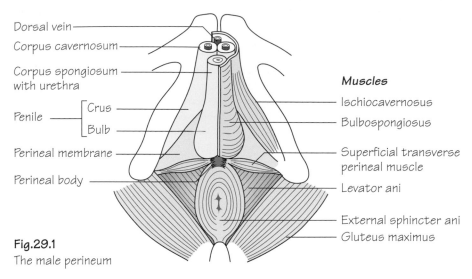

Dorsal vein
Corpus cavernosum
Corpus spongiosum with urethra
Penile — Crus / Bulb
Perineal membrane
Perineal body

Muscles
Ischiocavernosus
Bulbospongiosus
Superficial transverse perineal muscle
Levator ani
External sphincter ani
Gluteus maximus

Fig.29.1
The male perineum

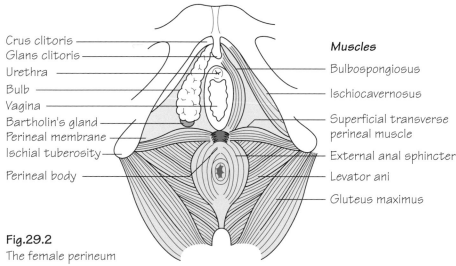

Crus clitoris
Glans clitoris
Urethra
Bulb
Vagina
Bartholin's gland
Perineal membrane
Ischial tuberosity
Perineal body

Muscles
Bulbospongiosus
Ischiocavernosus
Superficial transverse perineal muscle
External anal sphincter
Levator ani
Gluteus maximus

Fig.29.2
The female perineum

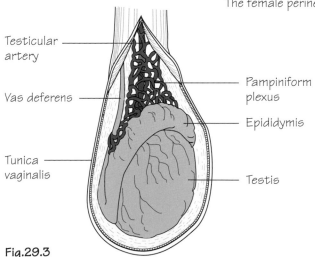

Testicular artery
Vas deferens
Tunica vaginalis

Pampiniform plexus
Epididymis
Testis

Fig.29.3
The testis and epididymis

The perineum lies below the pelvic diaphragm. It forms a diamond-shaped area when viewed from below that can be divided into an anterior *urogenital* region and a posterior *anal* region by a line joining the ischial tuberosities horizontally.

Anal region (Figs. 21.2 and 29.1)
The anal region contains the anal canal and ischiorectal fossae.
- **Anal canal**: is described earlier (p. 55).
- **Anal sphincter**: comprises external and internal components. The *internal anal sphincter* is a continuation of the inner circular smooth muscle of the rectum. The *external anal sphincter* is a skeletal muscular tube which, at its rectal end, blends with puborectalis to form an area of palpable thickening termed the *anorectal ring*. The competence of the latter is fundamental to anal continence.
- **Ischiorectal fossae**: lie on either side of the anal canal. The medial and lateral walls of the ischiorectal fossa are the levator ani and anal canal and the obturator internus, respectively. The fossae are filled with fat. The anococcygeal body separates the fossae posteriorly; however, infection in one fossa can spread anteriorly to the contralateral fossa

forming a horseshoe abscess. The *pudendal (Alcock's) canal* is a sheath in the lateral wall of the ischiorectal fossa. It conveys the pudendal nerve and internal pudendal vessels from the lesser sciatic notch to the deep perineal pouch (see below). The *inferior rectal branches* of the pudendal nerve and internal pudendal vessels course transversely across the fossa to reach the anus.

Urogenital region

The urogenital region is triangular in shape. The *perineal membrane* is a strong fascial layer that is attached to the sides of the urogenital triangle. In the male, it is pierced by the urethra and, in females, by the urethra and vagina.

In the female (Fig. 29.2)

• **Vulva**: is the term given to the female external genitalia. The *mons pubis* is the fatty protuberance overlying the pubic symphysis and pubic bones. The *labia majora* are fatty hair-bearing lips that extend posteriorly from the mons. The *labia minora* lie internal to the labia majora and unite posteriorly at the *fourchette*. Anteriorly, the labia minora form the *prepuce* and split to enclose the *clitoris*. The clitoris corresponds to the penis in the male. It has a similar structure in that it is made up of three masses of erectile tissue: the bulb (corresponding to the penile bulb) and right and left crura covered by similar but smaller muscles than those in the male. As in the male, these form the contents of the superficial perineal pouch. The deep perineal pouch, however, contains the vagina as well as part of the urethra, sphincter urethrae and internal pudendal vessels. The *vestibule* is the area enclosed by the labia minora, which contains the urethral and vaginal orifices. Deep to the posterior aspect of the labia majoris lie *Bartholin's glands* – a pair of mucus-secreting glands that drain anteriorly. They are not palpable in health but can become grossly inflamed when infected.

• **Urethra**: is short in the female (3–4 cm). It contributes to a relative predisposition to urinary tract infection in females from an upward spread of bowel organisms. The urethra extends from the bladder neck to the external meatus. The meatus lies between the clitoris and vagina.

• **Vagina**: measures approximately 8–12 cm in length. It is a muscular tube that passes upwards and backwards from the vaginal orifice. The cervix projects into the upper anterior aspect of the vagina creating *fornices* anteriorly, posteriorly and laterally. Lymph from the upper vagina drains into the internal and external iliac nodes. Lymph from the lower vagina drains to the superficial inguinal nodes. The blood supply to the vagina is from the *vaginal artery* (branch of the internal iliac artery) and the *vaginal branch of the uterine artery*.

In the male (Fig. 29.1)

The external urethral sphincter (striated muscle) lies deep to the perineal membrane within a fascial capsule termed the *deep perineal pouch*. In addition to the sphincter, two *glands of Cowper* are also contained within the deep pouch. The ducts from these glands pass forwards to drain into the bulbous urethra. Inferior to the perineal membrane is the *superficial perineal pouch*, which contains the:

• **Superficial transverse perineal muscles**: run from the perineal body to the ischial ramus.
• **Bulbospongiosus muscle**: covers the *corpus spongiosum*. The latter structure covers the spongy urethra.
• **Ischiocavernosus muscle**: arises on each side from the ischial ramus to cover the *corpus cavernosum*. It is the engorgement of venous sinuses within these cavernosa that generates and maintains an erection.

Hence, the penile root comprises a well-vascularised bulb and two crura which are supplied by branches of the internal pudendal artery. The erectile penile tissue is enclosed within a tubular fascial sheath. At the distal end of the penis the corpus spongiosum expands to form the *glans penis*. On the tip of the glans, the urethra opens as the *external urethral meatus*. The foreskin is attached to the glans below the meatus by a fold of skin – the *frenulum*.

The scrotum

The skin of the scrotum is thin, rugose and contains many sebaceous glands. A longitudinal *median raphe* is visible in the midline. Beneath the skin lies a thin layer of involuntary *dartos* muscle. The terminal spermatic cords, the testes and their epididymides are contained within the scrotum.

Testis and epididymis (Fig. 29.3)

The testes are responsible for spermatogenesis. Their descent to an extra-abdominal position favours optimal spermatogenesis as the ambient scrotal temperature is approximately 3°C lower than the body temperature.

• **Structure**: the testis is divided internally by a series of septa into approximately 200 lobules. Each lobule contains 1–3 *seminiferous tubules*, which anastomose into a plexus termed the *rete testis*. Each tubule is coiled when *in situ*, but when extended measures approximately 60 cm. *Efferent ducts* connect the rete testis to the epididymal head. They serve to transmit sperm from the testicle to the epididymis.
  • The *tunica vaginalis*, derived from the peritoneum, is a double covering into which the testis is invaginated.
  • The *tunica albuginea* is a tough fibrous capsule that covers the testis.
  • The *epididymis* lies along the posterolateral and superior borders of the testicle. The tunica vaginalis covers the epididymis with the exception of the posterior border.
  • The upper poles of both the testis and epididymis bear an *appendix testis* and *appendix epididymis* (*hydatid of Morgagni*), respectively.
• **Blood supply**: is from the testicular artery (a branch of the abdominal aorta, p. 43). Venous drainage from the testicle is to the pampiniform plexus of veins. The latter plexus lies within the spermatic cord, but coalesces to form a single vein at the internal ring. The left testicular vein drains to the left renal vein, whereas the right testicular vein drains directly to the inferior vena cava.
• **Lymphatic drainage**: is to the para-aortic lymph nodes.
• **Nerve supply**: is from T10 sympathetic fibres via the renal and aortic plexuses.

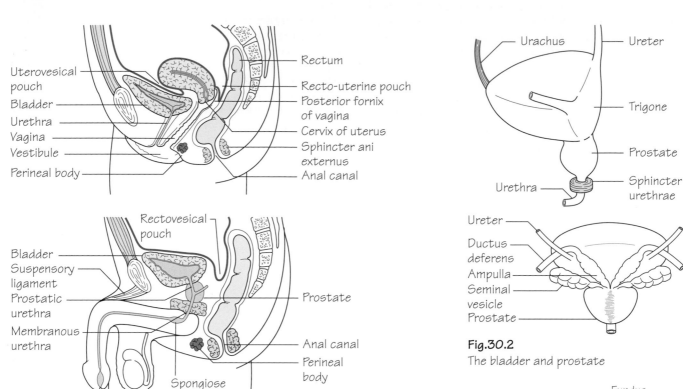

Fig.30.1
Sagittal sections through the male and female pelves

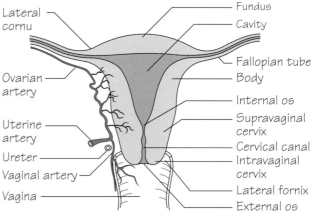

Fig.30.2
The bladder and prostate

Fig.30.3
A vertical section through the uterus and vagina.
Note the relation of the uterine artery to the ureter

Contents of the pelvic cavity (see Fig. 30.1)

- **Sigmoid colon:** see p. 55.
- **Rectum:** see p. 55.
- **Ureters:** see p. xx.
- **Bladder (Fig. 30.2):** see below.

Bladder

In adults, the bladder is a pelvic organ. It lies behind the pubis and is covered superiorly by peritoneum. It acts as a receptacle for urine and has a capacity of approximately 500 mL.

- **Structure:** the bladder is pyramidal in shape. The *apex* of the pyramid points forwards and from it a fibrous cord, the *urachus*, passes upwards to the umbilicus as the *median umbilical ligament*. The *base* (posterior surface) is triangular. In the male, the seminal vesicles lie on the outer posterior surface of the bladder and are separated by the vas deferens. The rectum lies behind. In the female, the vagina intervenes between the bladder and rectum. The *inferolateral* surfaces are related inferiorly to the pelvic floor and anteriorly to the retropubic fat pad and pubic bones. The bladder *neck* fuses with the prostate in the male, whereas it lies directly on the pelvic fascia in the female. The pelvic fascia is thickened in the form of the *puboprostatic ligaments* (male) and *pubovesical ligaments* to hold the bladder neck in position. The mucous membrane of the bladder is thrown into folds when the bladder is empty with the exception of the membrane overlying the base (termed the *trigone*), which is smooth. The superior angles of the trigone mark the openings of the ureteric orifices. A muscular elevation, the *interureteric ridge*, runs between the ureteric orifices. The inferior angle of the trigone corresponds to the *internal urethral meatus*. The muscle coat of the bladder is composed of a triple layer of trabeculated smooth muscle known as the *detrusor* (muscle). The detrusor is thickened at the bladder neck to form the *sphincter vesicae*.

- **Blood supply:** is from the superior and inferior vesical arteries (branches of the internal iliac artery, p. 69). The vesical veins coalesce around the bladder to form a plexus that drains into the internal iliac vein.

- **Lymph drainage:** is to the para-aortic nodes.

- **Nerve supply:** motor input to the detrusor muscle is from efferent parasympathetic fibres from S2–4. Fibres from the same source convey inhibitory fibres to the internal sphincter so that co-ordinated micturition can occur. Conversely, sympathetic efferent fibres inhibit the detrusor and stimulate the sphincter.

The male pelvic organs
The prostate (Fig. 30.2)
In health, the prostate is approximately the size of a walnut. It surrounds the prostatic urethra and lies between the bladder neck and the urogenital diaphragm. The apex of the prostate rests on the external urethral sphincter of the bladder. It is related anteriorly to the pubic symphysis but separated from it by extraperitoneal fat in the retropubic space (*cave of Retzius*). Posteriorly, the prostate is separated from the rectum by the *fascia of Denonvilliers*.
- **Structure**: the prostate comprises the anterior, posterior, middle and lateral lobes. On rectal examination, a posterior median groove can be palpated between the lateral lobes. The prostatic lobes contain numerous glands producing an alkaline secretion which is added to the seminal fluid at ejaculation. The prostatic glands open into the *prostatic sinus*. The *ejaculatory ducts*, which drain both the seminal vesicles and the vas, enter the upper part of the prostate and then the prostatic urethra at the *verumontanum*.
- **Blood supply**: is from the *inferior vesical artery* (branch of the internal iliac artery, p. 69). A prostatic plexus of veins is situated between the prostatic capsule and the outer fibrous sheath. The plexus receives the dorsal vein of the penis and drains into the internal iliac veins.

The vas deferens
The vas deferens conveys sperm from the epididymis to the ejaculatory duct from which it can be passed to the urethra. The vas arises from the tail of the epididymis and traverses the inguinal canal to the deep ring, passes downwards on the lateral wall of the pelvis almost to the ischial tuberosity and turns medially to reach the base of the bladder where it joins with the duct of the seminal vesicle to form the ejaculatory duct.

The seminal vesicles (Fig. 30.2)
The seminal vesicles consist of lobulated tubes which lie extraperitoneally on the bladder base lateral to the vas deferens.

The urethra (Fig. 30.1)
The male urethra is approximately 20 cm long (4 cm in the female). It is considered in three parts:
- **Prostatic urethra (3 cm)**: bears a longitudinal elevation (*urethral crest*) on its posterior wall. On either side of the crest a shallow depression, the *prostatic sinus*, marks the drainage point for 15–20 prostatic ducts. The *prostatic utricle* is a 5 mm blind-ending tract which opens into an eminence in the middle of the crest – the *verumontanum*. The ejaculatory ducts open on either side of the utricle.
- **Membranous urethra (2 cm)**: lies in the urogenital diaphragm and is surrounded by the external urethral sphincter (*sphincter urethrae*).
- **Penile urethra (15 cm)**: traverses the corpus spongiosum of the penis (see perineum, p. 71) to the external urethral meatus.

The female pelvic organs
The vagina
See perineum, p. 71.

The uterus and fallopian tubes (Fig. 30.3)
- **Structure**: the uterus measures approximately 8 cm in length in the adult nulliparous female. It comprises a *fundus* (part lying above the entrance of the fallopian tubes), *body* and *cervix*. The cervix is sunken into the anterior wall of the vagina and is consequently divided into *supravaginal* and *vaginal* parts. The internal cavity of the cervix communicates with the cavity of the body at the *internal os* and with the vagina at the *external os*. The fallopian tubes lie in the free edges of the broad ligaments and serve to transmit ova from the ovary to the cornua of the uterus. They comprise an *infundibulum*, *ampulla*, *isthmus* and *interstitial part*. The uterus is made up of a thick muscular wall (*myometrium*) and lined by a mucous membrane (*endometrium*). The endometrium undergoes massive cyclical change during menstruation.
- **Relations**: the uterus and cervix are related to the uterovesical pouch and superior surface of the bladder anteriorly. The recto-uterine pouch (of Douglas), which extends down as far as the posterior fornix of the vagina, is a posterior relation. The broad ligament is the main lateral relation of the uterus.
- **Position**: in the majority, the uterus is *anteverted*, i.e. the axis of the cervix is bent forward on the axis of the vagina. In some women, the uterus is *retroverted*.
- **Blood supply**: is predominantly from the uterine artery (a branch of the internal iliac artery, p. 69). It runs in the broad ligament and, at the level of the internal os, crosses the ureter at right angles to reach, and supply, the uterus before anastomosing with the ovarian artery (a branch of the abdominal aorta, p. 45).
- **Lymph drainage**: lymphatics from the fundus accompany the ovarian artery and drain into the para-aortic nodes. Lymphatics from the body and cervix drain to the internal and external iliac lymph nodes.

The ovary
Each ovary contains a number of primordial follicles, which develop in early fetal life and await full development into ova. In addition to the production of ova, the ovaries are also responsible for the production of sex hormones. Each ovary is surrounded by a fibrous capsule, the *tunica albuginea*.
- **Attachments**: the ovary lies next to the pelvic side wall and is secured in this position by two structures: the *broad ligament* which attaches the ovary posteriorly by the mesovarium; and the *ovarian ligament* which secures the ovary to the cornu of the uterus. The ovary is further attached to the *round ligament* which travserses the inguinal canal to attach to the labium majus.
- **Blood supply**: is from the ovarian artery (a branch of the abdominal aorta). Venous drainage is to the inferior vena cava on the right and to the left renal vein on the left.
- **Lymphatic drainage**: is to the para-aortic nodes.

31 Abdomen, developmental aspects

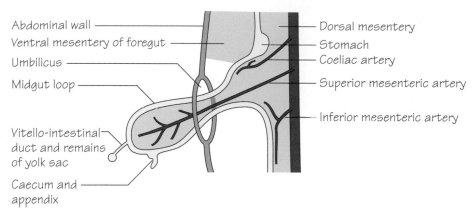

Abdominal wall
Ventral mesentery of foregut
Umbilicus
Midgut loop

Vitello-intestinal duct and remains of yolk sac

Caecum and appendix

Dorsal mesentery
Stomach
Coeliac artery
Superior mesenteric artery
Inferior mesenteric artery

Fig.31.1
The midgut loop

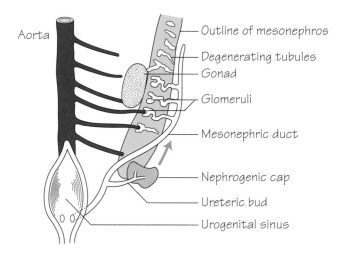

Aorta

Outline of mesonephros
Degenerating tubules
Gonad
Glomeruli
Mesonephric duct
Nephrogenic cap
Ureteric bud
Urogenital sinus

Fig.31.2
The mesonephros and related structures

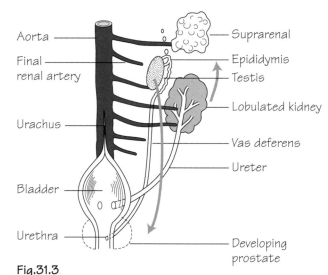

Aorta
Final renal artery
Urachus
Bladder
Urethra

Suprarenal
Epididymis
Testis
Lobulated kidney
Vas deferens
Ureter
Developing prostate

Fig.31.3
The upward migration of the kidney and the descent of the testis

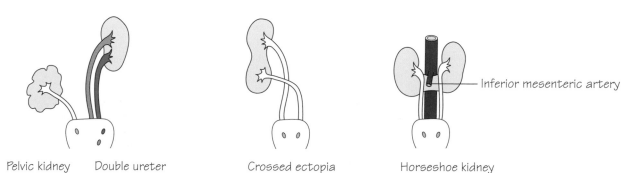

Pelvic kidney   Double ureter
Malrotation

Crossed ectopia

Horseshoe kidney

Inferior mesenteric artery

Fig.31.4
Anomalies of the kidney

The alimentary canal, development

The embryonic gut is suspended from the dorsal body wall by a continuous dorsal mesentery. The foregut also has a ventral mesentery. The *foregut* extends from the pharynx to the future entry of the common bile duct into the duodenum and receives its blood supply from the *coeliac artery*. The *midgut* ends about two-thirds of the way along the future transverse colon and is supplied by the *superior mesenteric artery*. During its development the midgut forms a loop which develops largely outside the body (Fig. 31.1), and it is not until 8–10 weeks of gestation that it returns to the abdomen. Attached to the apex of the loop is the *vitello-intestinal duct*, which connects it to the vestigeal *yolk sac*. The *hindgut* ends halfway down the future anal canal. It is supplied by the *inferior mesenteric artery*. The lower half of the anal canal is formed by an invagination of ectoderm and is supplied by the *inferior rectal artery*.

- **Rotation of the gut**: the return of the midgut loop to the abdomen is a complicated process which may be summed up as an anticlockwise rotation of the loop about the axis of the superior mesenteric artery. This throws the duodenum over to the right such that it lies in a 'C' shape with the duodenojejunal flexure to the left of the midline. The duodenum eventually loses its mesentery and becomes a retroperitoneal organ. As a result of rotation, the caecum comes to lie under the liver and then descends to its adult position in the right iliac fossa. The root of the small bowel mesentery is attached to the posterior abdominal wall between the duodenojejunal flexure and the ileocaecal junction. The ascending and descending colon also lose their mesenteries and become retroperitoneal.
- **The liver** develops as an outgrowth from the end of the foregut which divides into two parts. One division forms the gall bladder and cystic duct. The other part subdivides many times to produce the bile ducts and their subdivisions, together with the liver cells themselves.
- **The pancreas** develops as two outgrowths from the end of the foregut, and these subdivide and later fuse to form the pancreas and its two ducts.

The alimentary canal, developmental anomalies

- **Atresia**: is a congenital absence of continuity of the bowel. In the duodenum, this is thought to arise as a result of failure of recanalisation of the lumen following its occlusion by proliferation of the lining. In the small bowel and colon, atresia is thought to arise from a vascular insult to an area of the bowel during its development.
- **Gastroschisis and exomphalos**: failure of normal return of the midgut loop and an associated failure in development of the abdominal wall, result in the dramatic situation of an infant being delivered with part of the bowel on the outside of the abdomen. In gastroschisis the bowel is exposed, in exomphalos it is enclosed within a sac.
- **Meckel's diverticulum**: occasionally, the vitello-intestinal duct persists in whole or in part. The most common form of this anomaly is *Meckel's diverticulum*, which takes the form of a diverticulum from the terminal ileum. It is often said to be 2 inches long, 2 feet from the ileocaecal junction and present in 2% of cases, but these figures are very variable. It may contain ectopic gastric mucosa, which secretes acid. This may lead to bleeding or perforation. The diverticulum may be connected to the umbilicus by a fibrous cord, which may cause intestinal obstruction. Inflammation of a Meckel's diverticulum may mimic appendicitis. The distal part of the duct may persist to form a *raspberry tumour* at the umbilicus.
- **Malrotation of the gut**: partial failure of the normal process of rotation leads to a situation where the duodenum lies to the right of the midline and the caecum lies near the liver. The root of the small bowel

mesentery is, therefore, shorter than normal. This leaves the small bowel predisposed to rotation around the superior mesenteric artery. This is known as malrotation volvulus and, if not promply treated, leads to infarction of the entire small bowel.

- **Malformation of the bile ducts**: the cystic duct may run down alongside the common bile duct to join it in the lesser omentum or lower. The two hepatic ducts may similarly remain separate until lower down than usual. It is, therefore, important to indentify the anatomy of all the ducts precisely, before they are divided or opened.
- **Anorectal anomaly**: the hindgut may fail to join up with the ectodermal ingrowth so that the rectum ends blindly and a normal anus fails to form. This condition is frequently associated with a fistula from the rectum to the urinary tract.

The urogenital system, development

The embryonic 'kidney' is the *mesonephros*, a long ridge of excretory tissue on the dorsal body wall. Its tubules open into the *mesonephric duct*, which lies lateral to it and opens into the *urogenital sinus* which later becomes the bladder and urethra. The *testis* or *ovary* develops on the medial side of the mesonephros. The *ureteric bud* ascends from the lower end of the duct and picks up a *nephrogenic cap* from the lower end of the mesonephros, and this forms the glomeruli and renal tubules (Fig. 31.2). The collecting ducts, calices and ureters develop from the ureteric bud. The kidney ascends, being supplied by consecutive lateral branches of the aorta, until it reaches the suprarenal gland (Fig. 31.3). One of the aortic branches then enlarges to become the renal artery. The kidney finally rotates so that the hilum faces medially instead of anteriorly.

The upper mesonephric tubules are taken over by the testis and form the *efferent ducts*. A *paramesonephric duct* lies lateral to the mesonephric duct, crosses it and fuses with the opposite duct. They form the *uterine tubes*, the *uterus* and most of the *vagina*.

The lower end of the mesonephric duct is taken into the bladder, so that the ureteric opening assumes its adult position, and the rest of the mesonephric duct opens lower down into the urethra. It becomes the *vas deferens*.

The urogenital system, developmental anomalies (Fig. 31.4)

- **Duplex systems**: early division of the ureteric bud can lead to various varieties of double, or Y-shaped ureter.
- **Ectopia**: failure of the kidney to ascend results in a *pelvic kidney*. Occasionally, one or other kidney, usually the left, crosses the midline and fuses with the other kidney at its lower pole.
- **Aberrant renal arteries**: persistence of more than one of the lateral branches of the aorta may occur, so that aberrant vessels are present, usually entering at the lower pole. These may make transplantation of such a kidney difficult or impossible. Such arteries are *end-arteries* supplying a segment of the kidney, and their ligation will cause death of that part of the kidney.
- **Horseshoe kidney**: the lower ends of the two kidneys may fuse so that, when they ascend, the isthmus of the combined kidneys meets the inferior mesenteric artery which prevents further ascent. The two ureters cross the isthmus which may give rise to ureteric obstruction.
- **Autosomal dominant polycystic kidney**: this is due to localised dilatations of the tubules which give rise to a huge number of enormous fluid-containing cysts, giving the whole kidney a 'bubbling' appearance. Fifty per cent of patients go on to end-stage renal failure, and this condition accounts for about 10% of patients requiring renal dialysis and transplantation.

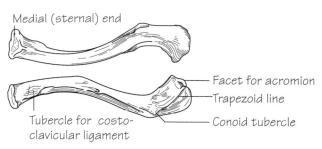

Medial (sternal) end

Facet for acromion

Trapezoid line

Conoid tubercle

Tubercle for costo-clavicular ligament

Fig.32.1
The upper and lower surfaces of the left clavicle

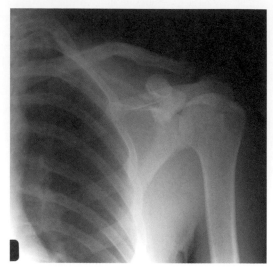

Fig.32.2
X-Ray of a fractured clavicle

The clavicle (Fig. 32.1)

• The clavicle is the first bone to ossify in the fetus (6 weeks).

• It develops in membrane and not in cartilage.

• Is subcutaneous throughout its length and transmits forces from the arm to the axial skeleton.

• The medial two-thirds are circular in cross-section and curved convex forwards. The lateral third is flat and curved convex backwards.

• The clavicle articulates medially with the sternum and 1st costal cartilage at the *sternoclavicular joint*. The bone is also attached medially to the 1st rib by strong *costoclavicular ligaments* and to the sternum by *sternoclavicular ligaments*.

• The clavicle articulates laterally with the acromion process of the scapula – the *acromioclavicular joint*. The *coracoclavicular ligaments* secure the clavicle inferolaterally to the coracoid process of the scapula. This ligament has two components – the *conoid* and *trapezoid ligaments*, which are attached to the *conoid tubercle* and *trapezoid line* of the clavicle, respectively.

• The clavicle is frequently fractured as a result of falling, particularly in riding accidents (Fig. 32.2 and see **Clinical notes**).

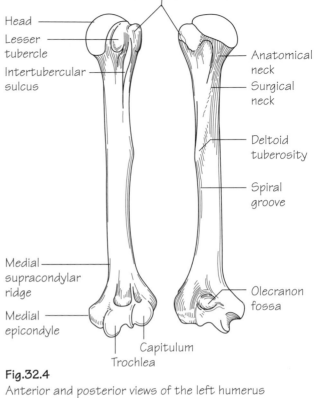

Fig.32.4
Anterior and posterior views of the left humerus

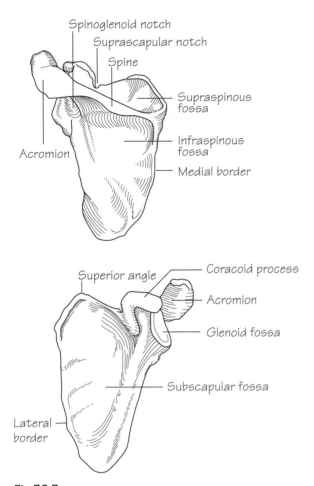

Fig.32.3
Posterior and anterior views of the left scapula

The scapula (Fig. 32.3)

• The scapula is triangular in shape. It provides an attachment for numerous muscles.
• The glenoid fossa articulates with the humeral head (*gleno-humeral joint*), and the acromion process with the clavicle (*acromioclavicular joint*).

The humerus (Fig. 32.4)

• The *humeral head* consists of one-third of a sphere. The rounded head articulates with the shallow *glenoid fossa*. This arrangement permits a wide range of shoulder movement.
• The *anatomical neck* separates the head from the greater and lesser tubercles. The *surgical neck* lies below the anatomical neck between the upper end of the humerus and shaft. The axillary nerve and circumflex vessels wind around the *surgical neck* of the humerus. These are at risk of injury in shoulder dislocations and humeral neck fractures (see Fig. 39.3).

• The *greater* and *lesser tubercles* provide attachment for the *rotator cuff* muscles. The tubercles are separated by the *intertubercular sulcus* in which the long head of the biceps tendon courses.
• A faint *spiral groove* is visible on the posterior aspect of the humeral shaft traversing obliquely downwards and laterally. The *medial* and *lateral heads of triceps* originate on either side of this groove. The radial nerve passes between the two heads.
• The *ulnar nerve* winds forwards in a groove behind the *medial epicondyle*.
• At the elbow joint, the *trochlea* articulates with the *trochlear notch of the ulna* and the rounded *capitulum* with the *radial head*. The medial border of the trochlea projects inferiorly a little further than the lateral border. This accounts for the *carrying angle*, i.e. the slight lateral angle made between the arm and forearm when the elbow is extended.

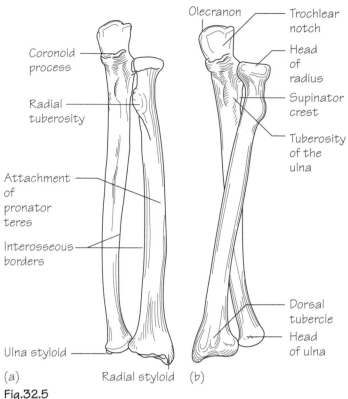

Coronoid process
Radial tuberosity
Attachment of pronator teres
Interosseous borders
Ulna styloid
(a)
Radial styloid

Olecranon
Trochlear notch
Head of radius
Supinator crest
Tuberosity of the ulna
Dorsal tubercle
Head of ulna
(b)

Fig.32.5
The left radius and ulna in (a) supination and (b) pronation

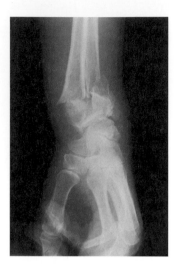

Fig.32.6
X-Ray of a fracture of the lower end of the radius (Colles' fracture)

The radius and ulna (Fig. 32.5)

• Both the radius and ulna have *interosseous*, *anterior* and *posterior* borders.

• The *biceps tendon* inserts into the roughened posterior part of the *radial tuberosity*. The anterior part of the tuberosity is smooth where it is covered by a bursa.

• The *radial head* is at its proximal end, whilst the *ulnar head* is at its distal end.

• The lower end of the radius articulates with the *scaphoid* and *lunate carpal bones* at the wrist joint. The distal ulna does not participate directly in the wrist joint.

• The *dorsal tubercle (of Lister)* is located on the posterior surface of the distal radius.

• In *pronation/supination* movements, the radial head rotates in the radial notch of the ulna and the radial shaft pivots around the relatively fixed ulna (connected by the interosseous ligament). The distal radius rotates around the head of the ulna.

The hand (Fig. 32.7)

The carpal bones are arranged into two rows. The palmar aspect of the carpus is concave. This is brought about by the shapes of the constituent bones and the *flexor retinaculum* bridging the bones anteriorly to form the *carpal tunnel* (see Fig. 43.1).

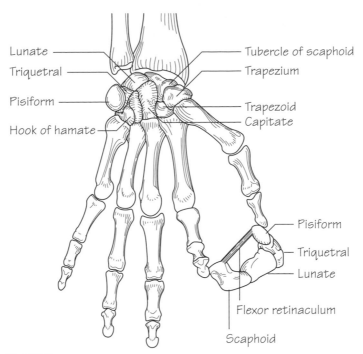

Lunate

Triquetral

Pisiform

Hook of hamate

Tubercle of scaphoid

Trapezium

Trapezoid

Capitate

Pisiform

Triquetral

Lunate

Flexor retinaculum

Scaphoid

Fig.32.7
The skeleton of the left hand, holding a cross-section
through the carpal tunnel

Clinical notes

- **Fracture of the clavicle (Fig. 32.2)**: the sternoclavicular joint is the only joint between the upper limb and the axial skeleton. Any force applied to the upper limb is transmitted from the gleno-humeral joint to the clavicle via the coracoclavicular ligament and, thence, via the clavicle to the sternum. Therefore, it is not surprising that the clavicle is one of the most frequently fractured bones in the body, usually as a result of falls on the tip of the shoulder or onto the outstretched hand. The weight of the upper limb drags the distal fragment downwards and the pull of the adductor muscles causes the fractured ends of the bone to overlap. The patient characteristically supports the weight of the injured limb with his or her other hand. Although the subclavian vessels and the brachial plexus lie immediately deep to the medial third of the clavicle, they are very rarely involved.
- **Fractures of the humerus**: three major nerves are closely related to the humerus and are liable to be damaged in fractures. Fractures of the surgical neck (or dislocation of the shoulder) may involve *the axillary nerve*; fractures of the shaft, the *radial nerve*; and fractures of the lower end that involve the medial epicondyle, the *ulnar nerve*. In any of these fractures, the functions of the corresponding nerves must be tested.
- **Colles' fracture**: this is a fracture of the lower end of the radius, usually caused by a fall onto the outstretched hand, and is frequently seen in elderly patients with osteoporosis. The fracture line is typically about 2.5 cm proximal to the wrist joint, and the distal fragment is displaced posteriorly and radially, giving the characteristic 'dinner-fork' deformity when seen from the side. Some degree of shortening also occurs as a result of impaction of the bone ends (Fig. 32.6).
- **Fractured scaphoid**: the scaphoid may be fractured by a fall on the outstretched hand. This injury is common in young adults and must be suspected clinically when tenderness is elicited by deep palpation in the anatomical snuffbox. Radiographic changes are often not apparent and, if effective treatment is not implemented, permanent wrist weakness and secondary osteoarthritis may follow. The blood supply to the scaphoid enters via its proximal and distal ends. However, in as many as one-third of cases, the blood supply enters only from the distal end. Under these circumstances, the proximal scaphoid fragment may be deprived of arterial supply and undergo avascular necrosis.

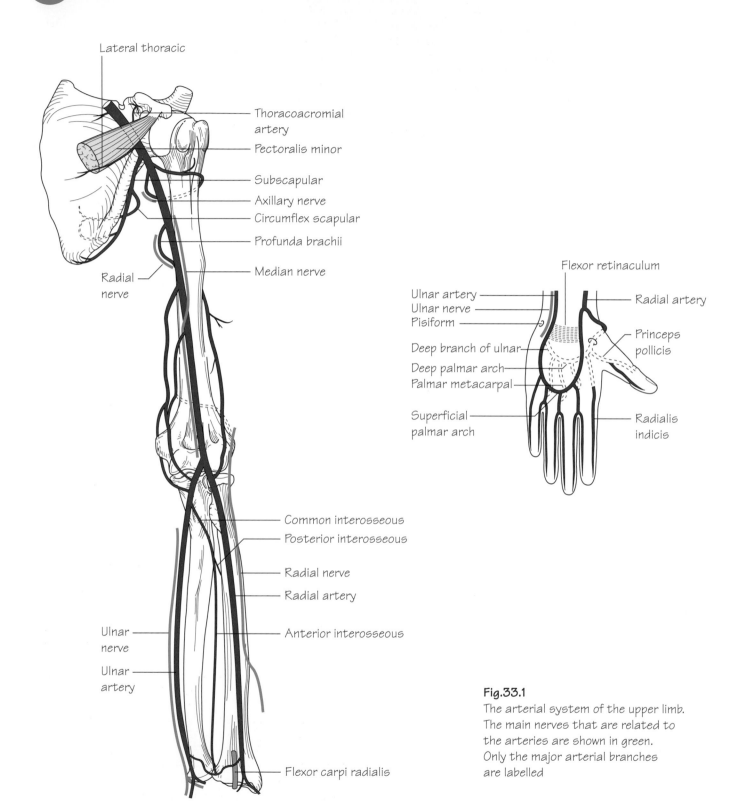

Lateral thoracic

Thoracoacromial artery

Pectoralis minor

Subscapular

Axillary nerve

Circumflex scapular

Profunda brachii

Median nerve

Radial nerve

Flexor retinaculum

Ulnar artery

Ulnar nerve

Pisiform

Deep branch of ulnar

Deep palmar arch

Palmar metacarpal

Superficial palmar arch

Radial artery

Princeps pollicis

Radialis indicis

Common interosseous

Posterior interosseous

Radial nerve

Radial artery

Ulnar nerve

Anterior interosseous

Ulnar artery

Flexor carpi radialis

Fig.33.1
The arterial system of the upper limb.
The main nerves that are related to
the arteries are shown in green.
Only the major arterial branches
are labelled

The axillary artery

- **Course**: the axillary artery commences at the lateral border of the 1st rib as a continuation of the subclavian artery (see Fig. 65.1) and ends at the *inferior border* of teres major where it continues as the *brachial artery*. The axillary vein is a medial relation throughout its course. The axillary artery is crossed anteriorly by *pectoralis minor*, which subdivides it into three parts (see Fig. 33.1). The first part has one branch, the second part two branches and the third part three branches:
  - *First part (medial to pectoralis minor)*: gives off the *superior thoracic artery*.
  - *Second part (behind pectoralis minor)*: gives off the *lateral thoracic artery* (which helps to supply the breast) and the *thoracoacromial artery* (p. 89).
  - *Third part (lateral to pectoralis minor)*: gives off the *subscapular artery*, which follows the lateral border of the scapula, and the *anterior and posterior circumflex humeral arteries*.

The brachial artery

- **Course**: the brachial artery commences at the inferior border of teres major as a continuation of the axillary artery and ends by bifurcating into the radial and ulnar arteries at the level of the neck of the radius. It lies immediately below the deep fascia throughout its course. The brachial artery is crossed superficially by the median nerve in the mid-arm from lateral to medial, and hence lies between the median nerve (medial relation) and biceps tendon (lateral relation) in the cubital fossa (see Fig. 41.3).
- **Branches**:
  - *Profunda brachii*: arises near the origin of the brachial artery and winds behind the humerus with the radial nerve in the spiral groove before taking part in the anastomosis around the elbow joint.
  - *Other branches*: include a *nutrient artery* to the humerus and *superior* and *inferior ulnar collateral* branches which ultimately take part in the anastomosis around the elbow.

The radial artery

- **Course**: the radial artery arises at the level of the neck of the radius from the bifurcation of the brachial artery. It passes over the biceps tendon to lie firstly on supinator and then descends on the radial side of the forearm, lying under the edge of brachioradialis in the upper half of its course and then between the tendons of brachioradialis and flexor carpi radialis in the lower forearm. The radial artery passes sequentially over supinator, pronator teres, the radial head of flexor digitorum superficialis, flexor pollicis longus and pronator quadratus. At the wrist, the artery lies on the distal radius lateral to the tendon of flexor carpi radialis. This is where the radial pulse is best felt.
- **Branches**:
  - *Palmar* and *dorsal carpal branches* are given off at the wrist.
  - A *superficial palmar branch* arises at the wrist, which supplies the thenar muscles and consequently anastomoses with the superficial palmar branch of the ulnar artery to form the *superficial palmar arch*.
  - The radial artery passes backwards under the tendons of abductor pollicis longus and extensor pollicis brevis to enter the *anatomical snuffbox*. It consequently passes over the scaphoid and trapezium in the snuffbox and exits by passing between the two heads of adductor pollicis to enter the palm and forms the *deep palmar arch* with a contribution from the ulnar artery (*deep palmar branch*). It gives off the *arteria princeps pollicis* to the thumb and the *arteria radialis indicis* to the index finger.
- The deep palmar arch gives off three *palmar metacarpal arteries*, which subsequently join the *common palmar digital arteries* (from the superficial arch) to supply the digits.

The ulnar artery

- **Course**: the ulnar artery commences at the bifurcation of the brachial artery at the level of the neck of the radius. It passes deep to the deep head of pronator teres and fibrous arch of flexor digitorum superficialis, before descending into the proximal part of the forearm, where it lies on flexor digitorum profundus and is overlapped by flexor carpi ulnaris. The ulnar nerve lies on its medial side throughout.

  At the wrist, both the ulnar artery and nerve lie lateral (radial) to flexor carpi ulnaris and pass *over* the flexor retinaculum giving carpal branches which contribute to the *dorsal* and *palmar carpal arches*.
- **Branches**:
  - A *deep palmar branch* completes the deep palmar arch (see above) and the ulnar artery continues as the *superficial palmar arch*, which is completed by the superficial palmar branch of the radial artery.
  - The *common interosseous artery* (see below).

The common interosseous artery

The common interosseous artery is the first ulnar branch to arise, and it subdivides into the:

- *Anterior interosseous artery*: descends with the interosseous branch of the median nerve on the anterior surface of the interosseous membrane. It predominantly supplies the flexor compartment of the forearm.
- *Posterior interosseous artery*: passes above the upper border of the interosseous membrane to enter the extensor compartment where it runs with the deep branch of the radial nerve supplying the extensor muscles of the forearm, eventually anastomosing with the anterior interosseous artery.

Clinical notes

- **Volkmann's ischaemic contracture**: is a deformity arising from ischaemia of the forearm muscles and subsequent healing by fibrosis. This causes contraction of the long flexor and extensor muscles. As the flexor muscles have greater bulk they contract to a greater degree than the extensors, leading to flexion of the wrist. The contraction of the long extensor muscles, which are inserted into the proximal phalanges, leads to extension of the metacarpophalangeal joints. The long flexors of the fingers are inserted into the middle and distal phalanges, so their contraction causes flexion of the interphalangeal joints.
- **The scapular anastomosis**: branches of the subclavian artery anastomose freely with branches of the axillary artery around the scapula (p. 149). The subscapular artery is usually the largest branch of the subclavian and, if the axillary artery is blocked or ligated *proximal* to the origin of the subscapular, blood may still reach the upper limb by reversal of flow in this vessel.

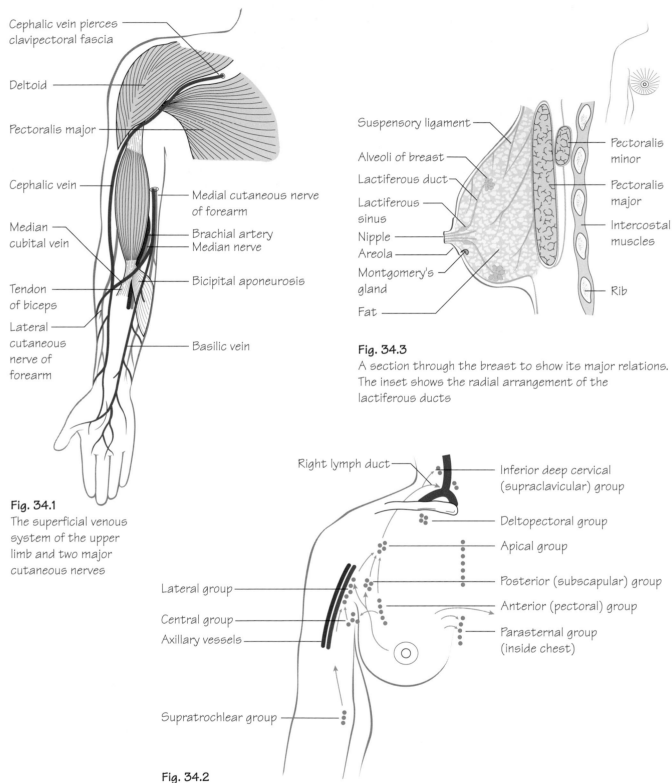

Fig. 34.1
The superficial venous system of the upper limb and two major cutaneous nerves

Labels (Fig. 34.1):
- Cephalic vein pierces clavipectoral fascia
- Deltoid
- Pectoralis major
- Cephalic vein
- Median cubital vein
- Tendon of biceps
- Lateral cutaneous nerve of forearm
- Medial cutaneous nerve of forearm
- Brachial artery
- Median nerve
- Bicipital aponeurosis
- Basilic vein

Fig. 34.3
A section through the breast to show its major relations. The inset shows the radial arrangement of the lactiferous ducts

Labels (Fig. 34.3):
- Suspensory ligament
- Alveoli of breast
- Lactiferous duct
- Lactiferous sinus
- Nipple
- Areola
- Montgomery's gland
- Fat
- Pectoralis minor
- Pectoralis major
- Intercostal muscles
- Rib

Fig. 34.2
The lymph nodes of the axilla and the lymphatic drainage of the breast

Labels (Fig. 34.2):
- Right lymph duct
- Inferior deep cervical (supraclavicular) group
- Deltopectoral group
- Apical group
- Posterior (subscapular) group
- Anterior (pectoral) group
- Parasternal group (inside chest)
- Lateral group
- Central group
- Axillary vessels
- Supratrochlear group

Venous drainage of the upper limb
(Fig. 34.1)

As in the lower limb, the venous drainage comprises interconnected superficial and deep systems.

- **The superficial system** comprises the *cephalic* and *basilic veins*.
  - The *cephalic vein* commences from the lateral end of the *dorsal venous network* overlying the anatomical snuffbox. It ascends the lateral, then anterolateral, aspects of the forearm and arm and finally courses in the deltopectoral groove to pierce the clavipectoral fascia and drain into the axillary vein.
  - The *basilic vein* commences from the medial end of the dorsal venous network. It ascends along the medial, then anteromedial, aspects of the forearm and arm to pierce the deep fascia (in the region of the mid-arm) to join with the *venae comitantes of the brachial artery* and form the *axillary vein*.

  The two superficial veins are usually connected by a median cubital vein in the cubital fossa.
- **The deep veins** consist of venae comitantes (veins which accompany arteries).

Lymphatic drainage of the chest wall and upper limb (Fig. 34.2)

Lymph from the chest wall and upper limb drains centrally via axillary, supratrochlear and infraclavicular lymph nodes.

Axillary lymph node groups

There are approximately 30–50 lymph nodes in the axilla. They are arranged into five groups:

- **Anterior (pectoral) group**: these lie along the anterior part of the medial wall of the axilla. They receive lymph from the upper anterior part of the trunk wall and breast.
- **Posterior (subscapular) group**: these lie along the posterior part of the medial wall of the axilla. They receive lymph from the upper posterior trunk wall down as far as the iliac crest.
- **Lateral group**: these lie immediately medial to the axillary vein. They receive lymph from the upper limb and the breast.
- **Central group**: these lie within the fat of the axilla. They receive lymph from all of the groups named above.
- **Apical group**: these lie in the apex of the axilla. They receive lymph from all of the groups mentioned above. From here, the lymph is passed to the *thoracic* duct (on the left) or *right lymphatic trunks* (see Fig. 5.3), with some passing to the *inferior deep cervical (supraclavicular) group*.

Lymph node groups in the arm

- The *supratrochlear group* of nodes lies subcutaneously above the medial epicondyle. They drain lymph from the ulnar side of the forearm and hand. Lymph from this group passes to the *lateral group* of axillary lymph nodes and thence drains centrally.
- A small amount of lymph from the radial side of the upper limb drains directly into the *infraclavicular group* of nodes. This group is arranged around the cephalic vein in the deltopectoral groove. From this point the efferent vessels pass through the clavipectoral fascia to drain into the *apical group* of axillary nodes and thence centrally. Thus, an infection of the thumb may cause enlargement of the infraclavicular nodes, whereas a lesion on the little finger may cause enlargement of the supratrochlear nodes.

The breast (Fig. 34.3)

The breasts are present in both sexes and have similar characteristics until puberty when, in the female, they enlarge and develop the capacity for milk production. The breasts are essentially specialised skin glands comprising fat, glandular and connective tissue. The base of the breast lies in a constant position on the anterior chest wall. It extends from the 2nd to 6th ribs anteriorly and from the lateral edge of the sternum to the mid-axillary line laterally. A part of the breast, the *axillary tail*, extends laterally through the deep fascia beneath pectoralis to enter the axilla. Each breast comprises 15–30 functional ducto-lobular units arranged radially around the nipple. The lobes are separated by fibrous septa (*suspensory ligaments*) which pass from the deep fascia to the overlying skin, thereby giving the breast structure. A lactiferous duct arises from each lobe and converges on the nipple. In its terminal portion, the duct is dilated (*lactiferous sinus*) and thence continues to the nipple from where milk can be expressed. The *areola* is the darkened skin that surrounds the nipple. Its surface is usually irregular due to multiple small tubercles – *Montgomery's glands*.

- **Blood supply**: is from the perforating branches of the internal thoracic artery (p. 21) and the lateral thoracic and thoracoacromial branches of the axillary artery (p. 81). The venous drainage corresponds to the arterial supply.
- **Lymphatic drainage**: from the lateral part of the breast to the anterior axillary nodes and, thence to the central and apical nodes. Lymph from the medial part of the breast drains into the internal mammary nodes (adjacent to the internal thoracic vessels beneath the chest wall), but variations in this pattern often occur.

Clinical notes

- **Superficial veins of the forearm**: these are of great clinical importance for phlebotomy and peripheral venous access. The most commonly used sites are the *median cubital vein* in the antecubital fossa and the *cephalic vein* in the forearm.
- **Dialysis**: a superficial vein can also be used repeatedly for *dialysis* for renal failure. An anastomosis is created between the radial artery and a superficial vein. As a result of the increased pressure, the vein becomes enlarged and the wall thickens (*arterialisation*). Two needles can then be inserted: the distal needle conveys blood to the dialyser and the blood is returned via the proximal needle. When dialysis is complete, the needles are removed.
- **Lymph drainage in carcinoma of the breast**: the axillary lymph nodes represent an early site of metastasis from primary breast malignancies, and their sampling provides important prognostic information as well as a basis for the choice of adjuvant treatment. Damage to axillary lymphatics during surgical clearance of axillary nodes or resulting from radiotherapy to the axilla increases the likelihood of subsequent upper limb lymphoedema.

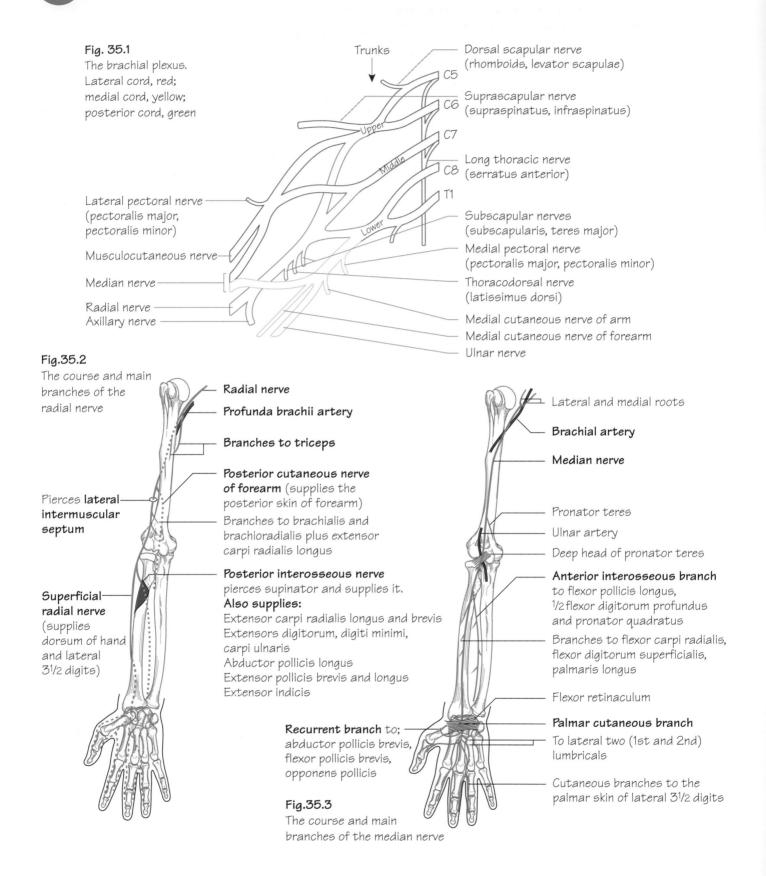

Fig. 35.1
The brachial plexus.
Lateral cord, red;
medial cord, yellow;
posterior cord, green

Trunks

C5

C6

C7

C8

T1

Upper

Middle

Lower

Dorsal scapular nerve
(rhomboids, levator scapulae)

Suprascapular nerve
(supraspinatus, infraspinatus)

Long thoracic nerve
(serratus anterior)

Subscapular nerves
(subscapularis, teres major)

Medial pectoral nerve
(pectoralis major, pectoralis minor)

Thoracodorsal nerve
(latissimus dorsi)

Medial cutaneous nerve of arm

Medial cutaneous nerve of forearm

Ulnar nerve

Lateral pectoral nerve
(pectoralis major,
pectoralis minor)

Musculocutaneous nerve

Median nerve

Radial nerve

Axillary nerve

Fig.35.2
The course and main
branches of the
radial nerve

Radial nerve

Profunda brachii artery

Branches to triceps

**Posterior cutaneous nerve
of forearm** (supplies the
posterior skin of forearm)

Branches to brachialis and
brachioradialis plus extensor
carpi radialis longus

Posterior interosseous nerve
pierces supinator and supplies it.
Also supplies:
Extensor carpi radialis longus and brevis
Extensors digitorum, digiti minimi,
carpi ulnaris
Abductor pollicis longus
Extensor pollicis brevis and longus
Extensor indicis

Pierces **lateral
intermuscular
septum**

**Superficial
radial nerve**
(supplies
dorsum of hand
and lateral
3½ digits)

Recurrent branch to;
abductor pollicis brevis,
flexor pollicis brevis,
opponens pollicis

Fig.35.3
The course and main
branches of the median nerve

Lateral and medial roots

Brachial artery

Median nerve

Pronator teres

Ulnar artery

Deep head of pronator teres

Anterior interosseous branch
to flexor pollicis longus,
½ flexor digitorum profundus
and pronator quadratus

Branches to flexor carpi radialis,
flexor digitorum superficialis,
palmaris longus

Flexor retinaculum

Palmar cutaneous branch

To lateral two (1st and 2nd)
lumbricals

Cutaneous branches to the
palmar skin of lateral 3½ digits

The brachial plexus (C5, C6, C7, C8 and T1) (Fig. 35.1)

- The plexus arises as five *roots*. These are the anterior primary rami of C5, C6, C7, C8 and T1. The roots lie between scalenus anterior and scalenus medius.
- The three *trunks* (upper, middle and lower) lie in the posterior triangle of the neck. They pass over the 1st rib to lie behind the clavicle.
- The *divisions* form behind the middle third of the clavicle around the axillary artery.
- The *cords* lie in the axilla and are related medially, laterally and posteriorly to the second part of the axillary artery.
- Terminal nerves arise from the cords surrounding the third part of the axillary artery.

The axillary nerve (C5 and C6)

- **Type**: mixed sensory and motor nerve.
- **Origin**: it arises from the *posterior cord* of the brachial plexus.
- **Course**: it passes through the quadrangular space (p. 89) with the posterior circumflex humeral artery. It provides: a motor supply to deltoid and teres minor; a sensory supply to the skin overlying deltoid; and an articular branch to the shoulder joint.

The radial nerve (C5, C6, C7, C8 and T1) (Fig. 35.2)

- **Type**: mixed sensory and motor.
- **Origin**: it arises as a continuation of the posterior cord of the brachial plexus.
- **Course and branches**: it runs with the profunda brachii artery between the long and medial heads of triceps into the posterior compartment and down between the medial and lateral heads of triceps. At the midpoint of the arm, it enters the anterior compartment by piercing the lateral intermuscular septum. In the region of the lateral epicondyle, the radial nerve lies under the cover of brachioradialis and divides into the superficial radial and posterior interosseous nerves.

The branches of the radial nerve include: *branches to triceps, brachioradialis* and *brachialis* as well as some cutaneous branches. The radial nerve terminates by dividing into two major nerves:

- The *posterior interosseous nerve*: passes between the two heads of supinator at a point three fingers' breadth distal to the radial head, thus passing into the posterior compartment. It supplies the extensor muscles of the forearm.
- The *superficial radial nerve*: descends the forearm under the cover of brachioradialis with the radial artery on its medial side. It terminates as cutaneous branches supplying the skin of the back of the wrist and hand.

The musculocutaneous nerve (C5, C6 and C7)

- **Type**: mixed sensory and motor.
- **Origin**: it arises from the lateral cord of the brachial plexus.
- **Course**: it passes laterally through the two conjoined heads of coracobrachialis and then descends the arm between brachialis and biceps, supplying all three of these muscles en route. It pierces the deep fascia just below the elbow (and becomes the *lateral cutaneous nerve of the forearm*). Here, it supplies the skin of the lateral forearm as far as the wrist.

The median nerve (C6, C7, C8 and T1) (Fig. 35.3)

- **Type**: mixed sensory and motor.
- **Origin**: it arises from the confluence of two roots from the medial and lateral cords lateral to the axillary artery in the axilla.
- **Course and branches**: the median nerve initially lies lateral to the brachial artery, but crosses it medially in the mid-arm. In the cubital fossa, it lies medial to the brachial artery which itself lies medial to the bicipital tendon. The median nerve then passes deep to the bicipital aponeurosis before passing between the two heads of pronator teres. A short distance below this the *anterior interosseous branch* is given off. This branch descends with the anterior interosseous artery to supply the deep muscles of the flexor compartment of the forearm, with the exception of the ulnar half of flexor digitorum profundus. In the forearm, the median nerve lies between flexor digitorum superficialis and flexor digitorum profundus and supplies the remaining flexors, except flexor carpi ulnaris. A short distance above the wrist it emerges from the lateral side of flexor digitorum superficialis and gives off the *palmar cutaneous branch*, which provides a sensory supply to the skin overlying the thenar eminence.

At the wrist, the median nerve passes beneath the flexor retinaculum in the midline and through the carpal tunnel into the palm, where it divides into its terminal branches: the *recurrent branch* to the muscles of the thenar eminence (but not adductor pollicis); the *branches to the 1st and 2nd lumbricals*; and the *cutaneous supply* to the palmar skin of the thumb, index, middle and lateral half of the ring fingers.

Clinical notes

- Brachial plexus injuries:
  - *Erb–Duchenne paralysis*: excessive downward traction on the upper limb during birth can result in injury to the C5 and C6 roots and the upper trunk of the brachial plexus. These nerves supply the abductors of the shoulder, the flexors of the elbow joint and the supinators so that, when these muscles are paralysed, the arm hangs down by the side with the forearm pronated and the palm facing backwards. This has been termed the 'waiter's tip' position.
  - *Klumpke's paralysis*: excessive upward traction on the upper limb can result in injury to the T1 root. As the latter is the nerve supply to the intrinsic muscles of the hand, this injury results in wasting of these muscles and clawing of the hand similar to that seen in ulnar nerve lesions. There is often an associated Horner's syndrome (ptosis, pupillary constriction and ipsilateral anhidrosis) as the traction injury often involves the cervical sympathetic chain also. Similar paralysis can occur when lesions in the region of the first rib compress the lower trunk of the brachial plexus, such as a tumour in the apex of the lung, or a cervical rib (p. 15).

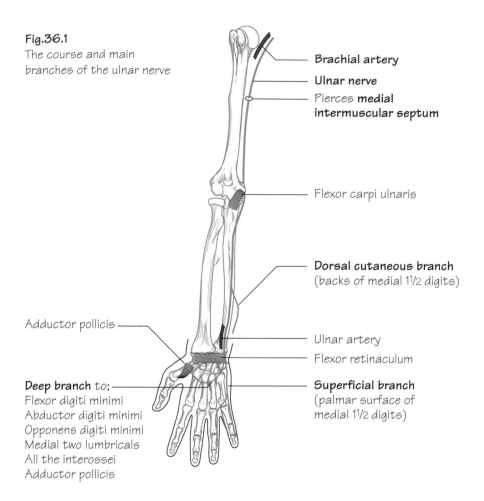

Fig.36.1
The course and main branches of the ulnar nerve

Brachial artery

Ulnar nerve

Pierces **medial intermuscular septum**

Flexor carpi ulnaris

Dorsal cutaneous branch
(backs of medial 1½ digits)

Adductor pollicis

Ulnar artery

Flexor retinaculum

Deep branch to:
Flexor digiti minimi
Abductor digiti minimi
Opponens digiti minimi
Medial two lumbricals
All the interossei
Adductor pollicis

Superficial branch
(palmar surface of medial 1½ digits)

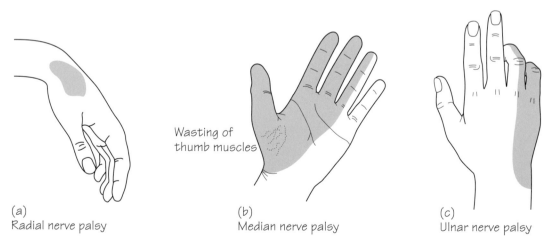

(a)
Radial nerve palsy

Wasting of thumb muscles

(b)
Median nerve palsy

(c)
Ulnar nerve palsy

Fig.36.2
Some common nerve palsies and the associated areas of altered sensation.
These are very variable

The ulnar nerve (C8 and T1) (Fig. 36.1)

- **Type**: mixed sensory and motor.
- **Origin**: from the medial cord of the brachial plexus.
- **Course and branches**: it runs on coracobrachialis to the mid-arm where it pierces the medial intermuscular septum with the superior ulnar collateral artery to enter the posterior compartment. It winds under the medial epicondyle and passes between the two heads of flexor carpi ulnaris to enter the forearm, where it innervates flexor carpi ulnaris and half of flexor digitorum profundus. In the lower forearm, the ulnar nerve lies medial to the ulnar artery and the tendon of flexor carpi ulnaris. *Dorsal* and *palmar cutaneous branches* are given off at this point. The ulnar nerve passes superficial to the flexor retinaculum and subsequently divides into terminal branches. These are:
  - The *superficial terminal branch*: terminates as terminal digital nerves innervating the skin of the little and medial half of the ring fingers.
  - The *deep terminal branch*: supplies the hypothenar muscles as well as two lumbricals, the interossei and adductor pollicis.

Other branches of the brachial plexus

Supraclavicular branches

- **Suprascapular nerve (C5 and C6)**: passes through the suprascapular notch to supply supra- and infraspinatus muscles.
- **Long thoracic nerve (of Bell) (C5, C6 and C7)**: supplies serratus anterior.

Infraclavicular branches

- **Medial and lateral pectoral nerves**: supply pectoralis major and minor.
- **Medial cutaneous nerves of the arm and forearm**.
- **Thoracodorsal nerve (C6, C7 and C8)**: supplies latissimus dorsi.
- **Upper and lower subscapular nerves**: supply subscapularis and teres major.

Clinical notes

Nerve injuries (see Fig. 36.2)

- **The axillary nerve**: is particularly prone to injury from the downward displacement of the humeral head during shoulder dislocations and by fracture of the surgical neck of the humerus.
  - *Motor deficit*: loss of deltoid abduction occurs with rapid wasting of this muscle. Loss of teres minor function is not detectable clinically.
  - *Sensory deficit*: is limited to the 'badge' region overlying the lower half of the deltoid groove.
- **The radial nerve**: may be injured in the spiral groove by fractures of the humeral shaft.
  - *Motor deficit*: weakness of all forearm extensors causes wrist-drop. Weakness of the grip occurs due to loss of the synergistic action of the extensors on the wrist (Chapter 42).
  - *Sensory deficit*: anaesthesia over the first dorsal web space (Fig. 36.2).
- **The median nerve (Fig. 36.2)**: is frequently compressed as it passes through the carpal tunnel (*carpal tunnel syndrome, see* Chapter 43). A lesion higher up the nerve may result from a supracondylar fracture of the humerus. This leads to:
  - *Motor deficit*: inability to flex the index and middle fingers and the distal phalanx of the thumb, including an ulnar deviation of the hand. Weakness and wasting of the thenar muscles occur, along with an inability to oppose and abduct thumb, with consequent impairment of the precision grip (p. 103).
  - *Sensory deficit*: anaesthesia over the radial 2/3 of the palm of the hand, thumb, index and middle fingers, as well as the radial half of the ring finger becomes anaesthetised.
- **The ulnar nerve (Fig. 36.2)**: injury occurs commonly at the elbow, causing high lesion (e.g. in fracture of the medial epicondyle). A low lesion commonly occurs as the ulnar nerve passes superficial to the flexor retinaculum at the wrist.
  - *Motor deficit*: with low lesions; the hand becomes 'clawed'. The interossei and lumbrical muscles normally produce flexion at the metacarpophalangeal joints and extension at the interphalangeal joints. Owing to the loss of interossei and lumbrical function, the reverse happens, and the metacarpophalangeal joints are extended and the interphalangeal joints are flexed. The deformity is less noticeable in the index and middle fingers because they retain their lumbrical muscles, which are supplied by the median nerve. When injury occurs at the elbow or above, the ring and little fingers get straighter as the ulnar supply to flexor digitorum profundus gets lost; however, this results in greater functional disability. The small muscles of the hand become wasted with the exception of the thenar and lateral two lumbrical muscles (all supplied by the median nerve). The wasting is best seen on the dorsum of the hand.
  - *Sensory deficit*: loss of sensation takes place over the ulnar 1/3 of the palm, the little finger and the ulnar half of the ring finger.

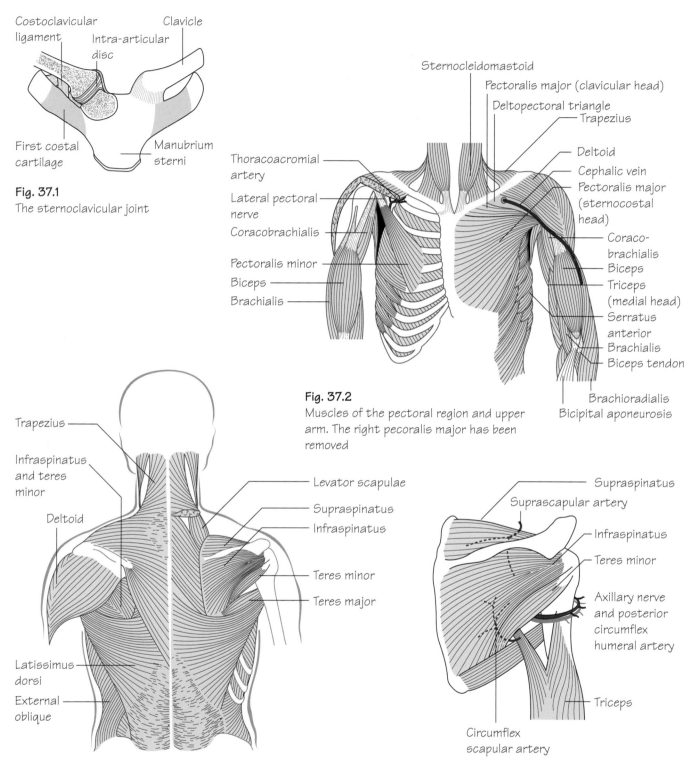

Fig. 37.1
The sternoclavicular joint

Costoclavicular ligament
Clavicle
Intra-articular disc
First costal cartilage
Manubrium sterni

Fig. 37.2
Muscles of the pectoral region and upper arm. The right pecoralis major has been removed

Sternocleidomastoid
Pectoralis major (clavicular head)
Deltopectoral triangle
Trapezius
Deltoid
Cephalic vein
Pectoralis major (sternocostal head)
Coraco-brachialis
Biceps
Triceps (medial head)
Serratus anterior
Brachialis
Biceps tendon
Brachioradialis
Bicipital aponeurosis
Thoracoacromial artery
Lateral pectoral nerve
Coracobrachialis
Pectoralis minor
Biceps
Brachialis

Fig. 37.3
Muscles of the scapular region and back. The right trapezius has been removed

Trapezius
Infraspinatus and teres minor
Deltoid
Latissimus dorsi
External oblique
Levator scapulae
Supraspinatus
Infraspinatus
Teres minor
Teres major

Fig. 37.4
The triangular and quadrangular spaces

Supraspinatus
Suprascapular artery
Infraspinatus
Teres minor
Axillary nerve and posterior circumflex humeral artery
Triceps
Circumflex scapular artery

Anatomy at a Glance, Third Edition. Omar Faiz, Simon Blackburn and David Moffat.

The upper limb is attached to the axial skeleton by the scapula and clavicle. It should be noted that this involves only the small *acromioclavicular* and *sternoclavicular joints*. The main attachment between the upper limb and the axial skeleton is muscular.

The muscles of the outer chest wall
(Figs. 37.2 and 37.3)
See Muscle index, pp. 178–179.
- Muscles of the outer anterior chest wall include the *pectoralis major* and *pectoralis minor*.
- Muscles of the back and shoulder include: *latissimus dorsi, trapezius, deltoid, levator scapulae, serratus anterior, teres major and minor, rhomboids major and minor, subscapularis, supraspinatus* and *infraspinatus*.

The sternoclavicular joint (Fig. 37.1)
- **Type:** atypical synovial joint.
- The articulation is between the sternal end of the clavicle and the manubrium. The articular surfaces are covered with fibrocartilage as opposed to the usual hyaline.
- A fibrocartilaginous articular disc separates the joint into two cavities.
- The fulcrum of movement at this joint is the costoclavicular ligament, thus, when the lateral end of the clavicle moves upwards, the medial end moves downwards.

The acromioclavicular joint
- **Type:** atypical synovial joint.
- The articulation is between the lateral end of the clavicle and the medial border of the acromion. As for the sternoclavicular joint, the articular surfaces are covered with fibrocartilage and an *articular disc* hangs into the joint from above.
- This is a weak joint. The main bond between the clavicle and the scapula is the *coracoclavicular ligament* (see Fig. 39.1).

The deltopectoral triangle, clavipectoral fascia and the anatomical spaces
(Fig. 37.2)
- The *deltopectoral triangle* is the region that is bounded by the deltoid, pectoralis major and, superiorly, by the small bare length of clavicle that does not provide attachment for either of these muscles.

- The *clavipectoral fascia* is a sheet of strong connective tissue. The uppermost part of this fascia forms the floor of the deltopectoral triangle. It is attached superiorly to the clavicle around the subclavius muscle. Below it splits to enclose pectoralis minor. The fascia continues downwards as the *suspensory ligament of the axilla* and becomes continuous with the fascial floor of the armpit. The clavipectoral fascia is pierced by four structures in total. Two structures drain inwards: (1) the *cephalic vein*; and (2) *lymphatics from the infraclavicular nodes*. Similarly, two structures pierce the fascia to pass outwards: (3) the *thoracoacromial artery*; and (4) the *lateral pectoral nerve* (which supplies pectoralis major and minor).
- Two important *anatomical spaces* are found in the shoulder region (Fig. 37.4):
  - The *quadrangular space* is an intermuscular space through which the axillary nerve and posterior circumflex humeral vessels pass through backwards to encircle the surgical neck of the humerus. It is bounded above by subscapularis and teres minor and below by teres major. The long head of triceps and the surgical neck of the humerus form the medial and lateral boundaries, respectively.
  - The *triangular space* is bound by teres major, teres minor and the long head of triceps. The circumflex scapular artery passes from front to back through this space to gain access to the infraspinous fossa.

Clinical notes
- **Sternal puncture**: red (haemopoietic) bone marrow is found in the adult only in the sternum, ilium, ribs, vertebral bodies, diploe of the flat skull bones and ends of some of the long bones. Between the sternal heads of pectoralis major the sternum is subcutaneous (Fig. 37.2), and this is, therefore, a convenient site for obtaining a sample of red bone marrow by the minor operation of *sternal puncture*.

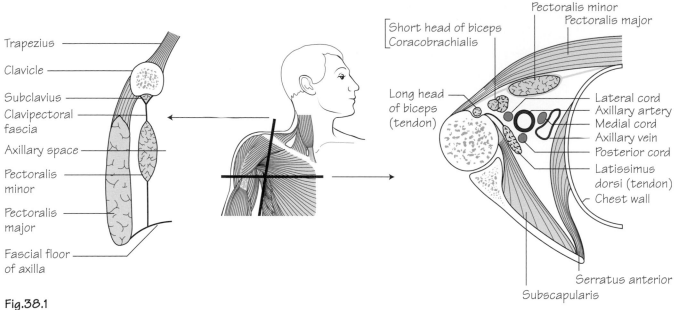

Fig.38.1
Vertical and horizontal sections through the axilla; the planes of the sections are shown in the central diagram

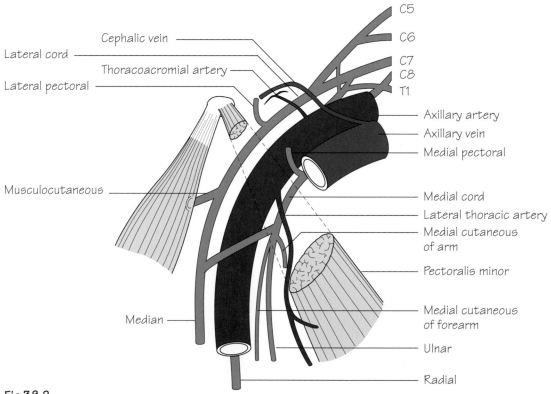

Fig.38.2
The main contents of the axilla from the front.
The posterior cord is hidden behind the axillary artery

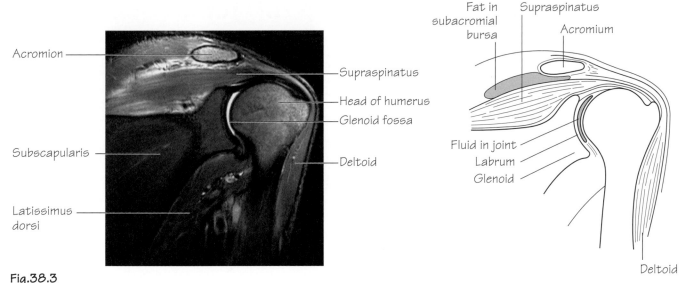

Fig.38.3

MRI showing normal coronal section of the left shoulder joint (see p. XX).
Note the close relation of supraspinatus to the acromion, with the underlying subacromial bursa

The major nerves and vessels supplying and draining the upper limb pass through the axilla.

The axilla is a three-sided pyramid. Its apex is the small region between the 1st rib, the clavicle and the scapula through which the major nerves and vessels pass.

The walls of the axilla are composed as follows:

• The *anterior wall* is made up from the pectoralis major, pectoralis minor and the clavipectoral fascia.

• The *posterior wall* is made up of the subscapularis, teres major and latissimus dorsi (Fig. 38.1).

• The *medial wall* consists of the upper part of serratus anterior, the upper ribs and intercostals.

• The *lateral wall* is very narrow and is formed by the tendon of latissimus dorsi as it inserts into the floor of the intertubercular (bicipital) groove. Close to this tendon, running downwards from above, are the coracobrachialis and short head of biceps, as well as the long head of biceps in the intertubercular groove.

The contents of the axilla (Figs. 38.1 and 38.2)

• **The axillary artery**: an important anastomosis exists between the subclavian artery and third part of the axillary artery – the *scapular anastomosis*. It compensates for compromised flow that may occur due to axillary artery obstruction. The principal arteries involved are the suprascapular, from the third part of the subclavian artery, and the subscapular, from the third part of the axillary artery.

• **The axillary vein**: is formed by the confluence of the venae comitantes of the axillary artery and the basilic vein (p. 83). It becomes the subclavian vein at the lateral border of the 1st rib. The named tributaries of the axillary vein correspond to those of the axillary artery.

• **The cords and branches of the brachial plexus**: see p. 84.

• **The axillary lymph nodes**: see p. 84.

• **Fat**.

Clinical notes

• **Axillary clearance**: in breast cancer surgery, the axillary lymph nodes are sometimes cleared. During the dissection for this procedure, one must clearly identify the axillary vein, the thoracodorsal nerve (C6, C7 and C8) and the long thoracic nerve (C5,C6 and C7). Injury to the thoracodorsal nerve results in paralysis of latissimus dorsi. Injury to the long thoracic nerve causes paralysis of serratus anterior, resulting in weakened arm abduction owing to the partial loss of the rotation of the scapula. This paralysis also causes *winging of the scapula*, which is demonstrated by asking the patient to lean against a wall with the weight taken on both hands. The affected scapula then protrudes backwards.

39 The shoulder (gleno-humeral) joint

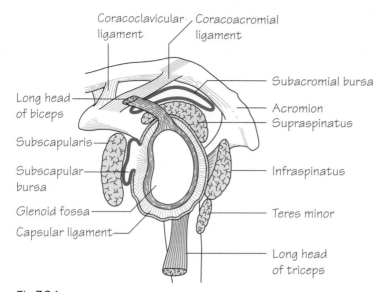

Fig.39.1
The glenoid cavity and its associated
ligaments and rotator cuff muscles

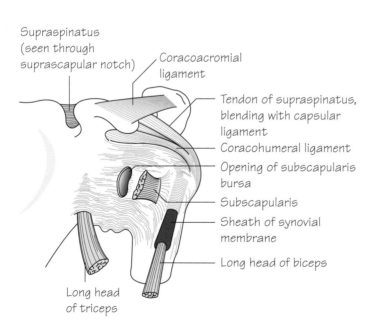

Fig.39.2
Anterior aspect of the shoulder joint

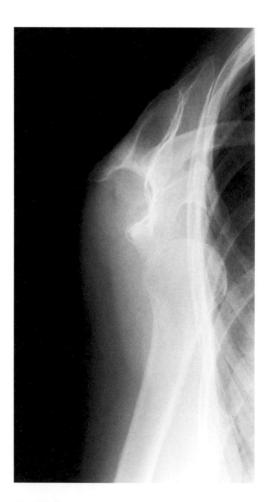

Fig.39.3
X-Ray of a dislocated shoulder

See Figs. 38.3, 39.1, and 39.2.

- **Type**: the shoulder is a *synovial 'ball and socket' joint* which permits multiaxial movement. It is formed by the articulation of the humeral head with the shallow glenoid fossa of the scapula (see p. 77). The glenoid is slightly deepened by a fibrocartilaginous rim – the *glenoid labrum*. Both articular surfaces are covered with hyaline cartilage.
- **The capsule**: of the shoulder joint is lax, permitting a wide range of movement. It is attached medially to the margins of the glenoid and laterally to the anatomical neck of the humerus, except inferiorly where it extends to the surgical neck. The capsule is significantly strengthened by slips from the surrounding rotator cuff muscle tendons.
- **Stability**: is afforded by the *rotator cuff* and the ligaments around the shoulder joint. The latter comprise: *three gleno-humeral ligaments*, which are weak reinforcements of the capsule anteriorly; a *coraco-humeral ligament*, which reinforces the capsule superiorly; and a *coraco-coacromial ligament*, which protects the joint superiorly. The main stability of the shoulder is afforded by the *rotator cuff*. The cuff comprises *subscapularis*, *supraspinatus* and, together, *infraspinatus* and *teres minor* (see Muscle index, p. 179), which pass in front of, above and behind the joint, respectively. Each of these muscles can perform its own function; when all are relaxed, free movement is possible but, when all are contracted, they massively reinforce shoulder stability.
- **Bursae**: two large bursae are associated with the shoulder joint. The *subscapular bursa* separates the shoulder capsule from the tendon of subscapularis which passes directly anterior to it. The subscapular bursa communicates with the shoulder joint. The *subacromial bursa* separates the shoulder capsule from the coracoacromial ligament above. The subacromial bursa does not communicate with the joint. The tendon of supraspinatus lies in the floor of the bursa.
- **The synovial membrane**: lines the capsule and covers the articular surfaces. It surrounds the intracapsular tendon of biceps and extends slightly beyond the transverse humeral ligament as a sheath. It forms the subscapular bursa anteriorly by protruding through the anterior wall of the capsule.
- **Nerve supply**: from the axillary (C5 and C6) and suprascapular (C5 and C6) nerves.

Shoulder movements

The shoulder is a 'ball and socket' joint allowing a wide range of movement. Much of this range is attributed to the articulation of the shallow glenoid with a rounded humeral head. The drawback, however, is that of compromised stability of the joint.

The principal muscles acting on the shoulder joint are:

- **Flexion (0–90°)**: pectoralis major, coracobrachialis and deltoid (anterior fibres).
- **Extension (0–45°)**: teres major, latissimus dorsi and deltoid (posterior fibres).
- **Internal (medial) rotators (0–40°)**: pectoralis major, latissimus dorsi, teres major, deltoid (anterior fibres) and subscapularis.
- **External (lateral) rotators (0–55°)**: infraspinatus, teres minor and deltoid (posterior fibres).
- **Adductors (0–45°)**: pectoralis major and latissimus dorsi.
- **Abductors (0–180°)**: supraspinatus, deltoid, trapezius and serratus anterior.

Although classified as an abductor, deltoid cannot start abduction when the arm is by the side as its fibres are more or less vertical. Abduction at the shoulder joint is, therefore, initiated by supraspinatus; deltoid continues it as soon as it obtains sufficient leverage. Almost simultaneously, the scapula is rotated so that the glenoid faces upwards; this action is produced by the lower fibres of serratus anterior, which are inserted into the inferior angle of the scapula, and by the trapezius, which pulls the lateral end of the spine of the scapula upwards and the medial end downwards.

Clinical notes

- **Shoulder dislocation (Fig. 39.3)**: as has been described above, stability of the shoulder joint is mostly afforded anteriorly, superiorly and posteriorly by the rotator cuff. Inferiorly, however, the shoulder joint is unsupported. Strong abduction and external rotation can, therefore, force the head of the humerus downwards and forwards to the point where the joint dislocates. This is termed anterior shoulder dislocation as the head usually comes to lie anteriorly in a subcoracoid position. The axillary nerve is sometimes damaged by this injury. The force of the injury may be sufficient to tear the glenoid labrum anteriorly, which facilitates recurrence. A surgical procedure is required when this tear leads to repeated dislocations.
- **Supraspinatus tendon rupture**: as supraspinatus is responsible for the initiation of abduction, traumatic rupture of its tendon prevents this movement unless the patient first leans over to the affected side, when abduction is carried out by gravity and deltoid can begin to work.
- **Painful arc syndrome**: inflammation of the supraspinatus tendon gives rise to pain when the shoulder is abducted between 60° and 120°. This is because the acromion impinges upon the inflamed supraspinatus tendon at this stage of abduction.

40 The arm

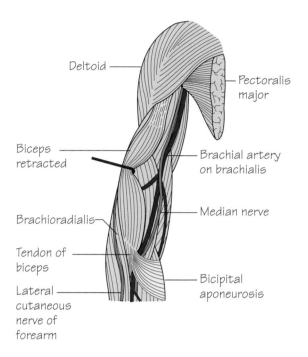

Deltoid

Pectoralis major

Biceps retracted

Brachial artery on brachialis

Median nerve

Brachioradialis

Tendon of biceps

Lateral cutaneous nerve of forearm

Bicipital aponeurosis

Fig.40.1
The main blood vessels and nerves
of the front of the arm

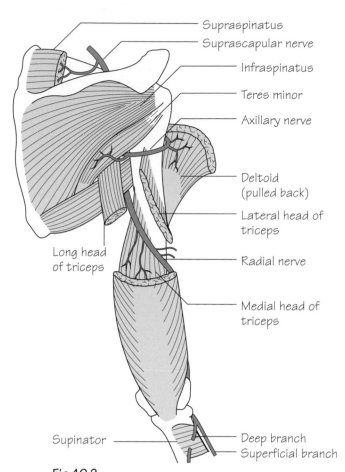

Supraspinatus

Suprascapular nerve

Infraspinatus

Teres minor

Axillary nerve

Deltoid (pulled back)

Lateral head of triceps

Radial nerve

Long head of triceps

Medial head of triceps

Supinator

Deep branch

Superficial branch

Fig.40.2
The major nerves in the back of the arm

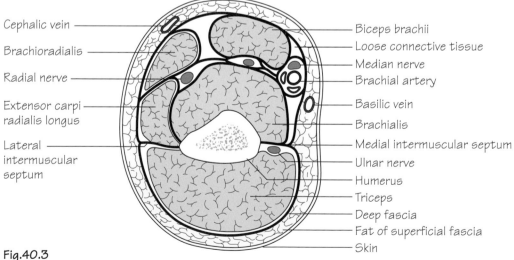

Cephalic vein

Brachioradialis

Radial nerve

Extensor carpi radialis longus

Lateral intermuscular septum

Biceps brachii

Loose connective tissue

Median nerve

Brachial artery

Basilic vein

Brachialis

Medial intermuscular septum

Ulnar nerve

Humerus

Triceps

Deep fascia

Fat of superficial fascia

Skin

Fig.40.3
Cross-section through the arm just above the elbow.
The thick black lines represent the deep fascia and the intermuscular septa

When viewed in cross-section, the arm consists of skin and subcutaneous tissue in which the superficial veins and sensory nerves course. Below lies a deep fascial layer. Medial and lateral intermuscular septa arise from the supracondylar lines of the humerus and extend to the deep fascia, thereby dividing the arm into anterior and posterior compartments.

- The anterior (flexor) compartment contents include (Figs. 40.1 and 40.3):
    - The *flexors of the elbow*: coracobrachialis, biceps and brachialis (see Muscle index, p. 179).
    - The *brachial artery* and its branches: see p. 81.
    - The *median nerve*: see p. 85.
    - The *ulnar nerve* in the upper arm only. The ulnar nerve pierces the medial intermuscular septum to pass into the posterior compartment in the mid-arm (p. 86).
    - The *musculocutaneous nerve and branches*. After providing motor innervation to the muscles of the flexor compartment, this nerve pierces the deep fascia in the mid-arm to become the lateral cutaneous nerve of the forearm (p. 85).

- The *basilic vein* in the upper arm only, as in the lower arm it is subcutaneous (p. 83).
- The posterior (extensor) compartment contents include (Figs. 40.2 and 40.3):
    - *Triceps*, the main extensor of the elbow (see Muscle index, p. 179).
    - The *radial nerve and branches*: see p. 85.
    - The *profunda brachii artery*: see p. 81.
    - The *ulnar nerve* in the lower arm, after it has pierced the medial intermuscular septum. (p. 87).

Clinical notes

- **Radial nerve injury**: the effects of damage to the radial nerve by fractures of the humerus are described in Chapter 36. Note, however, that some of the radial nerve branches to triceps arise in the axilla, so that triceps may be weakened but not completely paralysed by fractures of the humeral shaft. The associated sensory loss is small (see Fig. 36.2).

41 The elbow joint and cubital fossa

Attachment of capsular ligament
Coronoid fossa
Radial fossa
Trochlea
Capitulum
Coronoid process of ulna
Head of radius
Radial tuberosity

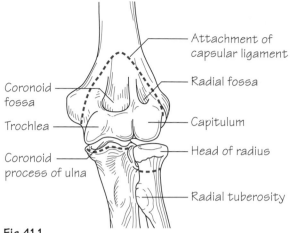

Fig.41.1
The bones of the elbow joint; the dotted lines represent the attachments of the capsular ligament

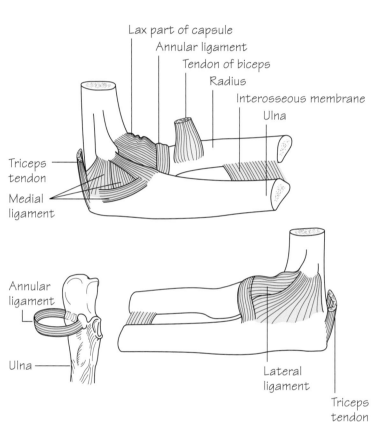

Lax part of capsule
Annular ligament
Tendon of biceps
Radius
Interosseous membrane
Ulna
Triceps tendon
Medial ligament
Annular ligament
Ulna
Lateral ligament
Triceps tendon

Fig.41.2
The ligaments of the elbow joint and the superior radio-ulnar joint

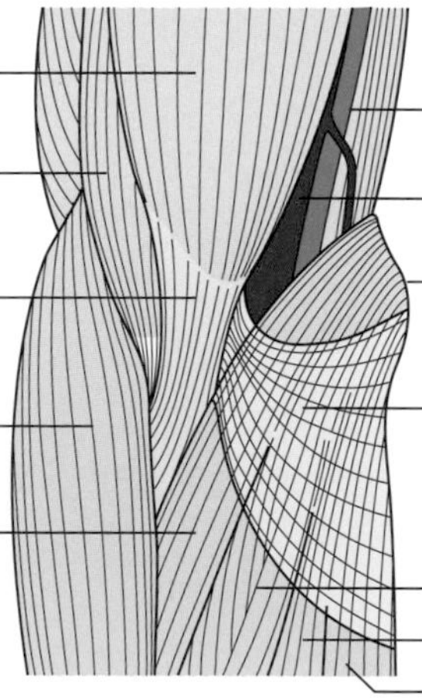

Biceps
Brachialis
Biceps tendon
Brachioradialis
Pronator teres
Median nerve
Brachial artery
Medial epicondyle
Bicipital aponeurosis
Flexor carpi radialis
Palmaris longus
Flexor carpi ulnaris

Fig.41.3
The cubital fossa. It is crossed by the median cubital vein

The elbow joint (Figs. 41.1 and 41.2)

• **Type**: *synovial hinge joint*. At the elbow, the humeral capitulum articulates with the radial head, and the trochlea of the humerus with the trochlear notch of the ulna. Fossae immediately above the trochlea and capitulum admit the coronoid process of the ulna and the radial head, respectively, during full flexion. Similarly, the olecranon fossa admits the olecranon process during full elbow extension. The elbow joint communicates with the superior radio-ulnar joint.

• **Capsule**: the capsule is lax in front and behind to permit full elbow flexion and extension. The non-articular medial and lateral epicondyles are extracapsular.

• **Ligaments** (Fig. 41.2): the capsule is strengthened medially and laterally by *collateral ligaments*.

  • The *medial collateral ligament* is triangular and consists of anterior, posterior and middle bands. It extends from the medial epicondyle of the humerus and the olecranon to the coronoid process of the ulna. The ulnar nerve is adjacent to the medial collateral ligament as it passes forwards below the medial epicondyle.

  • The *lateral collateral ligament* extends from the lateral epicondyle of the humerus to the *annular ligament*. The annular ligament is attached medially to the radial notch of the ulna and clasps, but does not attach to the radial head and neck. As the ligament is not attached to the head, this is free to rotate within the ligament.

The superior radio-ulnar joint

This is a *pivot joint*. It is formed by the articulation of the radial head and the radial notch of the ulna. The superior radio-ulnar joint communicates with the elbow joint.

Movements at the elbow

Flexion/extension occurs at the elbow joint. *Supination/pronation* occurs mostly at the *superior radio-ulnar joint* (in conjunction with movements at the inferior radio-ulnar joint).

• **Flexion (140°)**: biceps, brachialis, brachioradialis and the forearm flexor muscles.

• **Extension (0°)**: triceps and to a lesser extent anconeus.

• **Pronation (90°)**: pronator teres and pronator quadratus.

• **Supination (90°)**: biceps is the most powerful supinator. This movement is afforded by the insertion of the tendon of this muscle into the posterior aspect of the radial tuberosity. Supinator, extensor pollicis longus and brevis are weaker supinators.

The cubital fossa (Fig. 41.3)

• This fossa is defined by: a horizontal line joining the two epicondyles, the medial border of brachioradialis and the lateral border of pronator teres. The *floor* of the fossa consists of brachialis muscle and the overlying *roof* consists of superficial fascia. The median cubital vein runs within the roof of superficial fascia and connects the basilic and cephalic veins.

• Within the fossa the biceps tendon can be palpated. Medial to this lie the brachial artery and the median nerve.

• The radial and ulnar nerves lie outside the cubital fossa. The radial nerve passes anterior to the lateral epicondyle between brachialis and brachioradialis muscles. The ulnar nerve winds behind the medial epicondyle.

Clinical notes

• **Ulnar nerve damage**: owing to the close proximity of the ulnar nerve to the lower end of the humerus, it is at risk in many types of injury, e.g., fracture dislocations, compression and even surgical explorations (see Chapter 36 and Fig. 36.2).

• **Dislocation of the radial head**: the radial head may be pulled out of its annular ligament, particularly in children when an impatient adult suddenly pulls the hand of a reluctant child.

• **Dislocation of the elbow**: the classical injury is a posterior dislocation caused by a fall on the outstretched hand. It is most common in children whilst ossification is incomplete.

• **Superficial ulnar artery**: occasionally, the division of the brachial artery into radial and ulnar arteries takes place high in the arm and, when this occurs, the ulnar artery usually passes *superficial* to the flexor muscles of the forearm and may even be subcutaneous. If it is mistaken for a superficial vein, 'intravenous' injections may have catastrophic results.

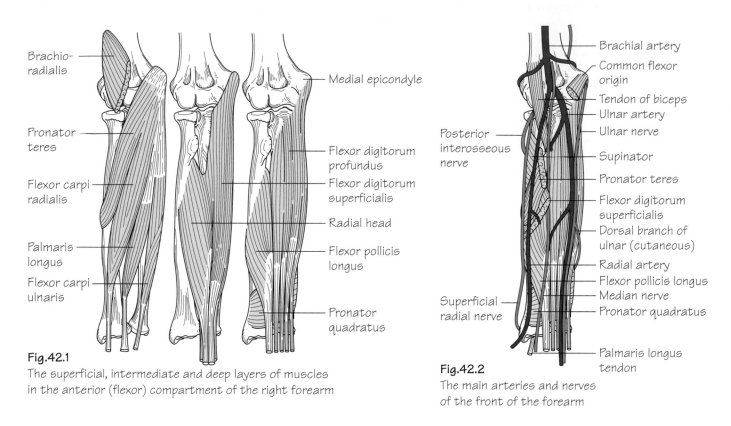

Brachio-
radialis

Pronator
teres

Flexor carpi
radialis

Palmaris
longus

Flexor carpi
ulnaris

Medial epicondyle

Flexor digitorum
profundus

Flexor digitorum
superficialis

Radial head

Flexor pollicis
longus

Pronator
quadratus

Fig.42.1
The superficial, intermediate and deep layers of muscles
in the anterior (flexor) compartment of the right forearm

Brachial artery

Common flexor
origin

Tendon of biceps

Ulnar artery

Ulnar nerve

Posterior
interosseous
nerve

Supinator

Pronator teres

Flexor digitorum
superficialis

Dorsal branch of
ulnar (cutaneous)

Radial artery

Flexor pollicis longus

Median nerve

Pronator quadratus

Superficial
radial nerve

Palmaris longus
tendon

Fig.42.2
The main arteries and nerves
of the front of the forearm

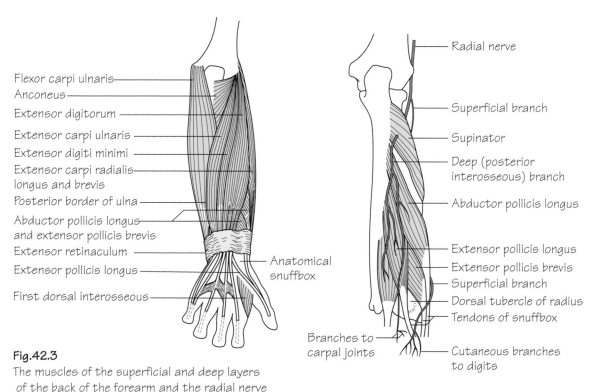

Flexor carpi ulnaris

Anconeus

Extensor digitorum

Extensor carpi ulnaris

Extensor digiti minimi

Extensor carpi radialis
longus and brevis

Posterior border of ulna

Abductor pollicis longus
and extensor pollicis brevis

Extensor retinaculum

Extensor pollicis longus

First dorsal interosseous

Anatomical
snuffbox

Fig.42.3
The muscles of the superficial and deep layers
of the back of the forearm and the radial nerve

Radial nerve

Superficial branch

Supinator

Deep (posterior
interosseous) branch

Abductor pollicis longus

Extensor pollicis longus

Extensor pollicis brevis

Superficial branch

Dorsal tubercle of radius

Tendons of snuffbox

Branches to
carpal joints

Cutaneous branches
to digits

The forearm is enclosed in deep fascia which is continuous with that of the arm. It is firmly attached to the periosteum of the subcutaneous border of the ulna. Together with the interosseous membrane, this divides the forearm into anterior and posterior compartments, each possessing its own muscles and arterial and nervous supplies. The superficial veins and cutaneous sensory nerves course in the subcutaneous tissue superficial to the deep fascia.

The interosseous membrane

• The interosseous membrane unites the interosseous borders of the radius and ulna. The fibres of this tough membrane run obliquely downwards and medially. A downward force (e.g. fall on the outstretched hand) is transmitted from the radius to the ulna and from here to the humerus and shoulder.

• The interosseous membrane provides attachment for neighbouring muscles.

The contents of the anterior (flexor) compartment of the forearm

• **Muscles (Fig. 42.1)**: the muscles within this compartment are considered in *superficial*, *intermediate* and *deep layers*. All of the muscles of the *superficial group* and part of *flexor digitorum superficialis* arise from the *common flexor origin* on the medial epicondyle of the humerus. With the exceptions of *flexor carpi ulnaris* and the *ulnar half of flexor digitorum profundus*, all of the muscles of the anterior compartment are supplied by the *median nerve* or its *anterior interosseous branch* (see Muscle index, p. 179).

• **Arteries (Fig. 42.2)**: ulnar artery and its anterior interosseous branch (via the common interosseous artery); radial artery.

• **Nerve supply (Fig. 42.2)**: median nerve and its anterior interosseous branch; ulnar nerve; superficial radial nerve.

The contents of the posterior fascial (extensor) compartment of the forearm

• **Muscles (Fig. 42.3)**: *brachioradialis* and *extensor carpi radialis longus* arise separately from the *lateral supracondylar ridge* of the humerus and are innervated by the main trunk of the *radial nerve*. The remaining extensor muscles are considered in *superficial* and *deep layers,* which are both innervated by the *posterior interosseous branch* of the radial nerve. The muscles of the superficial layer arise from the *common extensor origin* on the *lateral epicondyle of the humerus.* The muscles of the deep layer arise from the backs of the radius, ulna and interosseous membrane (see Muscle index, p. 180).

• **Arteries**: posterior interosseous artery (branch of the common interosseous artery).

• **Nerve supply**: posterior interosseous nerve (branch of the radial nerve) (Fig. 42.3).

Clinical notes

• **The power grip (see Chapter 44):** a lesion of the median nerve in the arm or forearm leads to paralysis of the long flexors except for those supplied by the ulnar nerve. This results in severe weakness in the power grip, for example in using a hammer. Lesions of the radial nerve also cause weakness of the power grip because the long flexors of the fingers also flex the wrist and the synergistic contraction of the extensors is necessary to prevent this. Other effects of median nerve lesions are considered in Chapter 36.

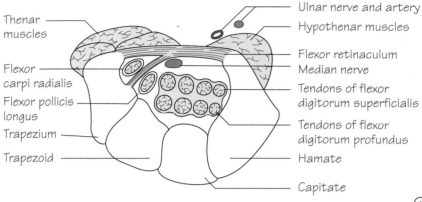

Thenar muscles
Flexor carpi radialis
Flexor pollicis longus
Trapezium
Trapezoid

Ulnar nerve and artery
Hypothenar muscles
Flexor retinaculum
Median nerve
Tendons of flexor digitorum superficialis
Tendons of flexor digitorum profundus
Hamate
Capitate

Fig.43.1
A diagrammatic cross-section through the carpal tunnel

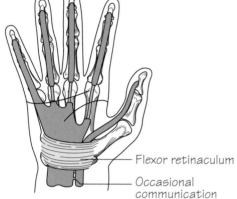

Flexor retinaculum
Occasional communication

Fig.43.2
The synovial sheaths of the flexor tendons

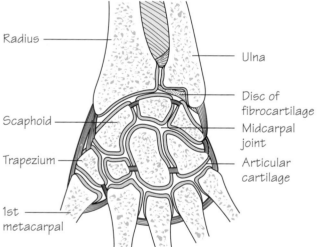

Radius
Scaphoid
Trapezium
1st metacarpal

Ulna
Disc of fibrocartilage
Midcarpal joint
Articular cartilage

Fig.43.3
The wrist and carpal joints

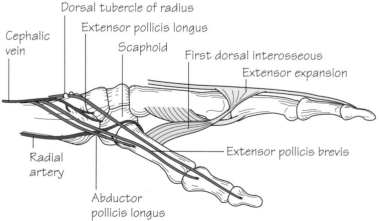

Cephalic vein
Dorsal tubercle of radius
Extensor pollicis longus
Scaphoid
First dorsal interosseous
Extensor expansion
Extensor pollicis brevis
Radial artery
Abductor pollicis longus

Fig.43.4
The anatomical snuffbox

The flexor retinaculum and carpal tunnel (Fig. 43.1)

The carpal tunnel is formed by the carpal bones and the overlying flexor retinaculum. It is through this tunnel that most, but not all, of the forearm tendons and the median nerve pass. The flexor retinaculum is attached to four bony points – the pisiform, the hook of the hamate, the scaphoid and the trapezium.

The carpal tunnel is narrow and no arteries or veins are transmitted through it for risk of potential compression. The median nerve is at risk of compression, however, when the tunnel is narrowed for any reason (see **Clinical notes**).

The synovial sheaths of the flexor tendons (Fig. 43.2)

The diagram illustrates the arrangement of the synovial sheaths that surround the flexor tendons. It can be seen that *flexor pollicis longus* has its own sheath (the *radial bursa*) and *flexor digitorum superficialis and profundus* share one (the *ulnar bursa*), which ends in the palm (except that for the little finger).

The wrist (radiocarpal) joint (Fig. 43.3)

• **Type**: the wrist is a *condyloid synovial joint*. The distal radius and a triangular disc of fibrocartilage covering the distal ulna form the proximal articulating surface. This disc is attached to the edge of the ulnar notch of the radius and to the base of the styloid process of the ulna and separates the wrist joint from the inferior radio-ulnar joint. The distal articulating surface is formed by the scaphoid and lunate bones with the triquetral participating in adduction.
• **Capsule**: a defined capsule surrounds the joint. It is thickened on either side by the *radial and ulnar collateral ligaments*.
• **Nerve supply**: from the anterior interosseous (median) and posterior interosseous (radial) nerves.

Wrist movements

Flexion/extension movements, occurring at the wrist, are accompanied by movements at the midcarpal joint. Of a total of 80° of wrist flexion, the majority occurs at the midcarpal joint, whereas in extension a corresponding increased amount occurs at the wrist joint.

The muscles acting on the wrist joint include:
• **Flexion**: all long muscles crossing the joint anteriorly.
• **Extension**: all long muscles crossing the joint posteriorly.
• **Abduction**: flexor carpi radialis and extensors carpi radialis longus and brevis.
• **Adduction**: flexor carpi ulnaris and extensor carpi ulnaris.

The joints of the hand (Fig. 43.3)

• **Intercarpal joints**: the *midcarpal joint*, located between the proximal and distal rows of carpal bones, is the most important of these as it participates in wrist movement (see above).

• **Carpometacarpal joints**: the most important of these is the *1st carpometacarpal (thumb) joint*. This is a saddle-shaped joint between the trapezium and the 1st metacarpal. It is a condyloid synovial joint which is separate from others in the hand, permitting a range of movement similar to that of a ball and socket joint. The most important movement of the thumb is opposition in which the thumb is opposed to the fingers, as in holding a pen.
• **Metacarpophalangeal joints**: are synovial condyloid joints.
• **Interphalangeal joints**: are synovial hinge joints.

The anatomical snuffbox

Fig. 43.4 illustrates the boundaries and contents of the anatomical snuffbox.

Clinical notes

• **Carpal tunnel syndrome**: the median nerve is at risk of compression as it passes through the confined space of the carpal tunnel. This sometimes occurs as a result of arthritis, fractures or swellings of adjacent structures, but more commonly occurs spontaneously. Compression gives rise to paraesthesiae and numbness in the thumb, index and middle fingers and part of the ring finger, and to weakness and wasting of the muscles of the thenar eminence. It can be relieved by medical treatment or by division of the flexor retinaculum.

• **Pulp space infections**: the pulp space of the tip of each finger is subdivided into compartments by strong fibrous septa which radiate from the distal phalanx. Infections thus lead to a rapid rise in pressure in these compartments causing severe pain. The septa must be broken down to achieve full drainage of pus resulting from such infections.

• **Tendon sheath infections**: infection of a tendon sheath leads to a painful swollen finger with limited movement and intense pain on passive extension. As can be seen from Fig. 43.2, infection of the little finger sheath can spread to the whole of the ulnar bursa and even to the radial bursa through a communication between them. Infections of the other fingers, however, are restricted to a single finger.

• **Carpal spaces**: these are potential spaces in the palm deep to the palmar aponeurosis. They are the *thenar space*, which surrounds the flexor tendons of the index finger and the thumb, and the *midpalmar space*, which contains the tendons of the other three fingers. They are separated by a fibrous septum, which passes from the deep surface of the palmar aponeurosis to the fascia covering adductor pollicis. Infections of the spaces, which are uncommon, can give rise to misleading signs, as the swelling associated with the infection affects the loose tissues on the dorsum of the hand even though the infection is in the palm.

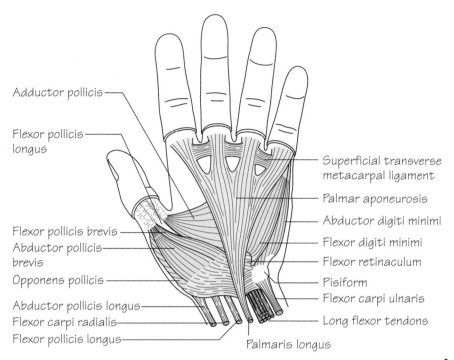

Adductor pollicis

Flexor pollicis longus

Flexor pollicis brevis

Abductor pollicis brevis

Opponens pollicis

Abductor pollicis longus

Flexor carpi radialis

Flexor pollicis longus

Palmaris longus

Superficial transverse metacarpal ligament

Palmar aponeurosis

Abductor digiti minimi

Flexor digiti minimi

Flexor retinaculum

Pisiform

Flexor carpi ulnaris

Long flexor tendons

Fig.44.1
The superficial muscles of the hand

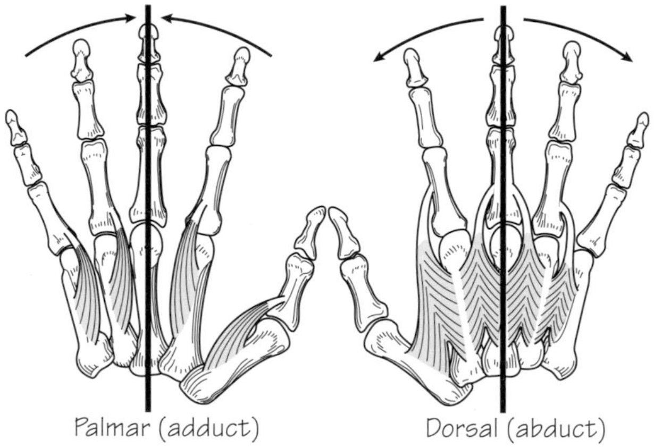

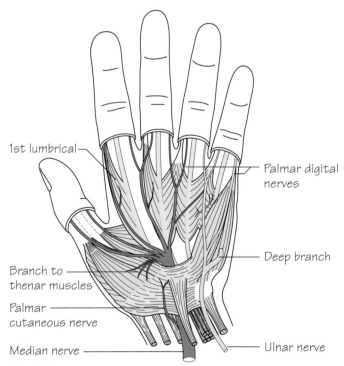

1st lumbrical

Palmar digital nerves

Branch to thenar muscles

Palmar cutaneous nerve

Median nerve

Deep branch

Ulnar nerve

Fig.44.3
The ulnar (yellow) and median (green) nerves in the hand.
Note particularly the recurrent branch of the median nerve which supplies the thenar muscles

Palmar (adduct)   Dorsal (abduct)

Fig.44.2
The palmar (L.) and dorsal (R.) interossei of the left hand and their actions in abduction and adduction

Anatomy at a Glance, Third Edition. Omar Faiz, Simon Blackburn and David Moffat.

The palm of the hand (Fig. 44.1)

- **Skin**: the skin of the palm is bound to underlying fascia by fibrous bands.
- **Deep fascia**: the *palmar aponeurosis* is a triangular layer which is attached to the distal border of the flexor retinaculum. Distally, the aponeurosis splits into four slips at the bases of the fingers which blend with the fibrous flexor sheaths (see below). The aponeurosis provides firm attachment of the overlying skin with protection of the underlying structures.
- **Fibrous flexor sheaths**: these are fibrous tunnels in which the flexor tendons and their synovial sheaths lie. They arise from the metacarpal heads and pass to the bases of the distal phalanges on the anterior aspect of the digits. They insert into the sides of the phalanges. These sheaths are lax over the joints and thick over the phalanges and, hence, do not restrict flexion.
- **Synovial flexor sheaths**: these are sheaths that limit friction between the flexor tendons and the carpal tunnel and fibrous flexor sheaths.
- **Long flexor tendons**: the tendons of flexor digitorum superficialis (FDS) divide into two halves at the level of the proximal phalanx and pass around flexor digitorum profundus (FDP), where they reunite. At this point they, then, split again to insert into the sides of the middle phalanx. FDP continues along its path to insert into the distal phalanx. Flexor pollicis longus (FPL) passes through the carpal tunnel in its own synovial sheath and inserts into the distal phalanx. The tendons of flexor carpi radialis, palmaris longus and flexor carpi ulnaris pass through the forearm and also insert in the proximal hand (see Muscle index, p. 179).

Muscles of the hand (Fig. 44.1)

- The *thenar muscles*: these are the short muscles of the thumb. They include abductor pollicis brevis, flexor pollicis brevis, opponens pollicis and adductor pollicis.
- The *hypothenar muscles*: these are the short muscles of the little finger. They include abductor digiti minimi, flexor digiti minimi and opponens digiti minimi.
- *Lumbricals*: these four muscles arise from the tendons of FDP. They insert into the radial side of each of the proximal phalanges and into the dorsal extensor expansions. The lumbricals serve to flex the metacarpophalangeal joints while extending the interphalangeal joints.
- The *interosseous muscles* (Fig. 44.2): these comprise eight muscles which arise from the shafts of the metacarpals. They are responsible for flexion at the metacarpophalangeal joints and extension of the interphalangeal joints. They perform abduction and adduction movements of the fingers. These movements occur around the middle finger; hence, adduction is the bringing together of all fingers towards the middle finger; abduction is moving them away from the middle finger. The dorsal interossei each arise from two metacarpals and insert into the proximal phalanges so as to provide abduction (D.AB). Abduction of the little finger is achieved by abductor digiti minimi. The palmar interossei arise from only one metacarpal and are inserted into the proximal phalanges so as to provide adduction (P.AD). Note that the middle finger cannot be adducted (and hence has no palmar interosseous), but can be abducted in either direction and so has two dorsal interosseous insertions.

The dorsum of the hand

- **Skin**: unlike the palm of the hand, the skin is thin and freely mobile over the underlying tendons.
- **Long extensor tendons**: the four tendons of extensor digitorum (ED) pass under the extensor retinaculum. On the dorsum of the hand, the ED tendon to the index finger is accompanied by the tendon of extensor indicis. The ED tendon to the little finger is accompanied by the double tendon of extensor digiti minimi. The ED tendons of the little, ring and middle fingers are connected to each other by fibrous slips. On the posterior surface of each finger, the extensor tendon spreads to form a dorsal digital expansion. This expansion is triangular shaped and, at its apex, splits into three parts: a middle slip which is attached to the base of the middle phalanx; and two lateral slips which converge to attach to the base of the distal phalanx. The base of the expansion receives the appropriate interossei and lumbricals. The tendons of abductor pollicis longus and extensor pollicis brevis and longus form the boundaries of the anatomical snuffbox and proceed to insert into the thumb.

Neurovascular structures of the hand (Fig. 44.3)

See chapters on upper limb: arteries, nerves, veins and lymphatics.

Movements of the fingers and thumb

The hand is required to perform a versatile range of movements including, in particular, two types of grip. These are:

- **The power grip**: this utilizes the whole hand and the grip is used, for example, for holding a hammer or squeezing a rubber ball. It is carried out by the long flexor tendons, aided by the contraction of the wrist extensors (p. 99).
- **The precision grip**: this is the grip used for holding a pair of forceps or threading a needle. It involves flexion at the metacarpophalangeal joints and extension of the interphalangeal joints of the fingers and opposition of the thumb. These movements are carried out by the interossei and lumbricals of the fingers and the opponens pollicis and other muscles of the thenar eminence, respectively.

Clinical notes

- **Preservation of the thumb**: the thumb is by far the most important of the digits because of its ability to oppose the other fingers. For this reason, the thumb must be preserved at all costs, even when it is severely injured. Various ingenious operations have been devised to replace a thumb which has been irretrievably damaged.
- **Testing the median nerve**: the flexor pollicis brevis is *usually* supplied not only by the median nerve but also by a branch from the ulnar. The opponens pollicis *often* receives a branch from the ulnar. For this reason, the abductor pollicis brevis is used to test for the integrity of the median nerve. The patient is asked to move his or her thumb, against resistance, away from the plane of the palm of the hand.

45 Surface anatomy of the upper limb

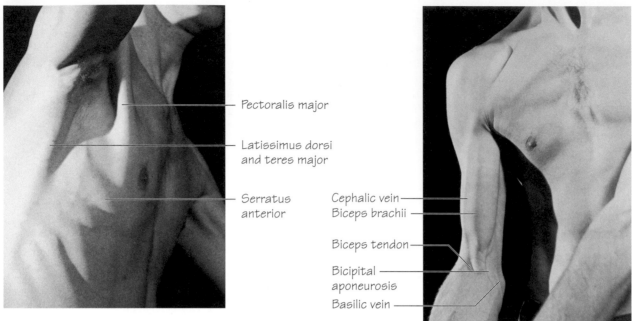

Pectoralis major
Latissimus dorsi and teres major
Serratus anterior

Fig.45.1
The axilla with the arm fully abducted

Cephalic vein
Biceps brachii
Biceps tendon
Bicipital aponeurosis
Basilic vein

Fig.45.2
The biceps tendon and aponeurosis which are a guide to the positions of the brachial artery and the median nerve at the elbow

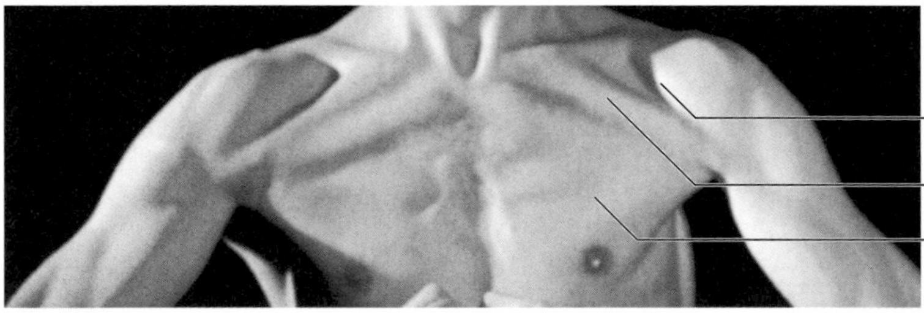

Deltopectoral triangle
Clavicular head
Sternocostal head
of pectoralis major

Fig.45.3
Strong contraction of the pectoral muscles produced by adduction

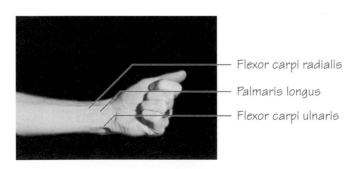

Flexor carpi radialis
Palmaris longus
Flexor carpi ulnaris

Fig.45.4
The visible tendons at the front of the wrist.
Palmaris longus is a guide to the position of the median nerve

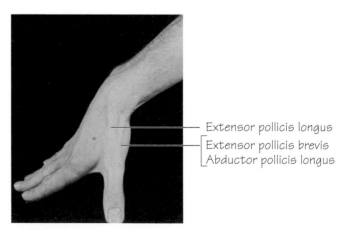

Extensor pollicis longus
Extensor pollicis brevis
Abductor pollicis longus

Fig.45.5
The anatomical snuffbox.
Details are shown in Fig.43.4

 Anatomy at a Glance, Third Edition. Omar Faiz, Simon Blackburn and David Moffat.

Bones and joints

- **Vertebrae**: if a finger is passed down the posterior neck in the midline, the first bony structure palpated is the spinous process of the 7th cervical vertebra (*vertebra prominens*) – the first six spinous processes being covered by the *ligamentum nuchae*.
- **Scapula**: the *acromion process* can be palpated as a lateral extension of the *spine of the scapula*. The *spine, superior angle, inferior angle* and *medial border* are palpable posteriorly. The *coracoid process* can be palpated below the clavicle anteriorly within the lateral part of the deltopectoral triangle (Fig. 45.3).
- **Clavicle**: is subcutaneous and, therefore, palpable throughout its length.
- **Humerus**: the *head* is palpable in the axilla with the shoulder abducted. The *lesser tuberosity* can be felt lateral to the *coracoid process*. When the arm is externally and internally rotated, the lesser tuberosity can be felt moving next to the fixed coracoid process.
- **Elbow**: the *medial* and *lateral epicondyles of the humerus* and *olecranon process of the ulna* can be palpated in line when the elbow is extended. With the elbow flexed, they form a triangle. This assumes importance clinically in differentiating supracondylar fractures of the humerus, where the 'triangle' is preserved, from elbow dislocations, where the olecranon comes into line with the epicondyles.
- **Radius**: the *radial head* can be felt in a hollow, distal to the lateral epicondyle on the posterolateral aspect of the extended elbow. The head can be felt rotating when the forearm is pronated and supinated.
- **Ulna**: the posterior border is subcutaneous and therefore palpable.
- **Wrist**: the *styloid processes of the radius and ulna* are palpable. The *dorsal tubercle (of Lister)* can be felt on the posterior aspect of the distal radius.
- **Hand**: the *pisiform* can be palpated at the base of the hypothenar eminence. The *hook of the hamate* can be felt on deep palpation in the hypothenar eminence just distal to the pisiform. The *scaphoid bone* can be felt within the anatomical snuffbox (Fig. 45.5).

The soft tissues

- **Axilla**: the anterior axillary fold (formed by the lateral border of pectoralis major) and the posterior axillary fold (formed by latissimus dorsi as it passes around the lower border of teres major) are easily palpable (Fig. 45.1).
- **Pectoralis major**: contracts strongly during arm adduction (Fig. 45.3); this is useful in the examination of breast lumps.
- **Breast**: the base of the breast overlaps the 2nd to 6th ribs and extends from the sternum to the mid-axillary line. The nipple (in males) usually overlies the 4th intercostal space.

- **Anterior wrist**: the *proximal transverse crease* corresponds to the level of the wrist joint. The *distal transverse crease* lies at the level of the proximal border of the flexor retinaculum.
- **Anatomical snuffbox**: the boundaries are formed medially by extensor pollicis longus and laterally by the tendons of abductor pollicis longus and extensor pollicis brevis.

Vessels

- The *subclavian artery* can be felt pulsating as it crosses the 1st rib.
- The *brachial artery* bifurcates into radial and ulnar branches at the level of the neck of the radius. The brachial pulse is felt by pressing laterally at a point medial to the bicipital tendon (Fig. 45.2). This is the pulse used when taking blood pressure measurements.
- At the wrist, the *radial artery* courses on the radial side of flexor carpi radialis (Fig. 45.4) and the *ulnar artery* and nerve course on the radial side of flexor carpi ulnaris. The pulses of both are easily felt at these points. The radial artery can also be felt in the anatomical snuffbox.
- The *superficial palmar arch* is impalpable and reaches as far as the proximal palmar crease. The *deep palmar arch* reaches a point approximately one finger's breadth proximal to the superficial arch.
- The *dorsal venous network* (on the dorsum of the hand) drains laterally into the cephalic vein and medially into the basilic vein. These veins can be identified in most lean subjects. The median cubital vein is usually visible in the cubital fossa.

Nerves

The *ulnar nerve* can usually be rolled as it courses behind the medial epicondyle – an important point when considering surgical approaches to the elbow and fractures of the medial epicondyle.

The surface markings of *impalpable nerves* must be known for safe surgical incisions:

- **Axillary nerve**: winds around behind the surgical neck of the humerus.
- **Radial nerve**: crosses from medial to lateral behind the midpoint of the humeral shaft.
- **Posterior interosseous branch (of radial nerve)**: winds around the radius, three fingers' breadth distal to the head of the radius.
- **Median nerve (at the wrist)**: lies in the midline, just lateral to the tendon of palmaris longus.
- **Ulnar nerve (at the wrist)**: lies immediately medial to the ulnar artery.

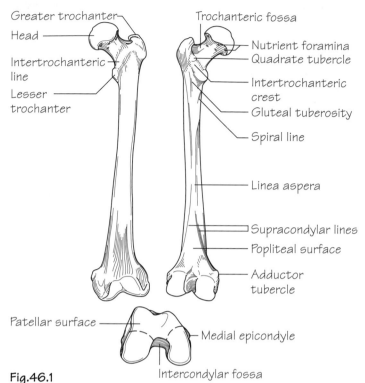

Fig.46.1
The left femur, anterior and posterior views
and the lower end from below

Labels: Greater trochanter, Head, Intertrochanteric line, Lesser trochanter, Trochanteric fossa, Nutrient foramina, Quadrate tubercle, Intertrochanteric crest, Gluteal tuberosity, Spiral line, Linea aspera, Supracondylar lines, Popliteal surface, Adductor tubercle, Patellar surface, Medial epicondyle, Intercondylar fossa

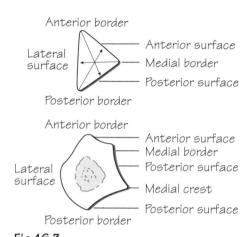

Fig.46.3
Diagram to explain the borders and surfaces
of the fibula (see text)

Labels: Anterior border, Lateral surface, Posterior border, Anterior surface, Medial border, Posterior surface, Anterior border, Lateral surface, Posterior border, Anterior surface, Medial border, Posterior surface, Medial crest, Posterior surface

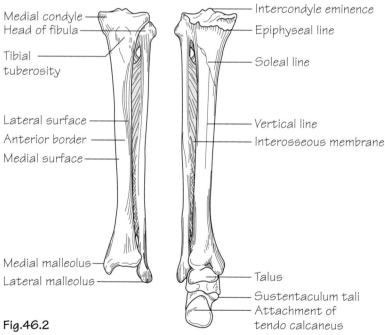

Fig.46.2
The front and back of the left tibia, fibula and ankle region.
The interosseous membrane and its openings are also shown

Labels: Medial condyle, Head of fibula, Tibial tuberosity, Lateral surface, Anterior border, Medial surface, Medial malleolus, Lateral malleolus, Intercondyle eminence, Epiphyseal line, Soleal line, Vertical line, Interosseous membrane, Talus, Sustentaculum tali, Attachment of tendo calcaneus

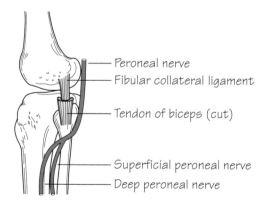

Fig.46.4
The left knee viewed from the lateral side
showing the common peroneal nerve

Labels: Peroneal nerve, Fibular collateral ligament, Tendon of biceps (cut), Superficial peroneal nerve, Deep peroneal nerve

The femur (Fig. 46.1)

The femur is the longest bone in the body. It has the following characteristic features:

• The *femoral head* articulates with the *acetabulum* of the hip bone at the hip joint. It extends from the femoral neck and is rounded, smooth and covered with articular cartilage. This configuration permits a wide range of movement. The head faces medially, upwards and forwards into the acetabulum. The *fovea* is the central depression on the head to which the *ligamentum teres* is attached.

• The *femoral neck* forms an angle of 125° with the femoral shaft. Pathological lessening or widening of the angle is termed *coxa vara* and *coxa valga* deformity, respectively.

• The *femoral shaft* constitutes the length of the bone. At its upper end it carries the *greater trochanter* and, posteromedially, the *lesser trochanter*. Anteriorly, the rough *trochanteric line* and, posteriorly, the smooth *trochanteric crest* demarcate the junction between the shaft and neck. The *linea aspera* is the crest seen running longitudinally along the posterior surface of the femur splitting in the lower portion into the *supracondylar lines*. The medial supracondylar line terminates at the *adductor tubercle*.

• The lower end of the femur comprises the medial and lateral *femoral condyles*. These bear the articular surfaces for articulation with the tibia at the knee joint. The lateral condyle is more prominent than the medial. This prevents lateral displacement of the patella. The condyles are separated posteriorly by a deep *intercondylar notch*. Anteriorly, the lower femoral aspect is smooth for articulation with the posterior surface of the patella.

The tibia (Fig. 46.2)

The tibia serves to transfer weight from the femur to the talus. It has the following characteristics:

• The flattened upper end of the tibia – the *tibial plateau* – comprises medial and lateral *tibial condyles* for articulation with the respective femoral condyles. In contrast to the femoral condyles, the medial tibial condyle is the larger of the two.

• The *intercondylar area* is the space between the tibial condyles on which can be seen two projections – the *medial* and *lateral intercondylar tubercles*. Together these constitute the *intercondylar eminence*. The horns of the lateral meniscus are attached close to either side of the eminence.

• On the anterior upper shaft the *tibial tuberosity* is easily identifiable. This is the site of insertion of the *ligamentum patellae*.

• The shaft is triangular in cross-section. It has anterior, medial and lateral borders and posterior, lateral and medial surfaces.

• The anterior border and medial surface of the shaft are subcutaneous throughout its length. For this reason, the tibial shaft is the most common site for open fractures.

• On the posterior surface of the shaft an oblique line – the *soleal line* – demarcates the tibial origin of soleus. Popliteus inserts into the triangular area above the soleal line.

• The fibula articulates with the tibia superiorly at an articular facet on the postero-inferior aspect of the lateral condyle – the *superior tibiofibular joint* (*synovial*).

• The *fibular notch* is situated laterally on the lower end of the tibia for articulation with the fibula at the *inferior tibiofibular joint* (*fibrous*).

• The tibia projects inferiorly as the *medial malleolus*. It constitutes the medial part of the mortice that stabilises the talus. The medial malleolus is grooved posteriorly for the passage of the tendon of tibialis posterior.

The fibula (Fig. 46.2)

The fibula does *not* form part of the knee joint and does *not* participate in weight transmission. The main functions of the fibula are to provide origin for muscles and to participate in the ankle joint. It has the following characteristic features:

• The *styloid process* is a prominence on the fibular *head* onto which the tendon of biceps is inserted (around the *lateral collateral ligament*) (Fig. 46.4).

• The fibular *neck* separates the head from the fibular shaft. The common peroneal nerve is a close relation as it winds around the neck prior to dividing into superficial and deep branches (Fig. 46.4).

• The fibula is triangular in cross-section. It has anterior, medial (interosseous) and posterior borders with anterior, lateral and posterior surfaces. The *medial crest* is on the posterior surface (Fig. 46.3).

• The lower end of the fibula is the *lateral malleolus*. This is the lateral part of the mortice that stabilises the talus. It bears a smooth medial surface for articulation with the talus. The posterior aspect of the malleolus is grooved for the passage of the tendons of peroneus longus and brevis. The lateral malleolus projects further downwards than the medial malleolus.

The patella

• The ligamentum patellae, which is attached to the apex of the patella and the tibial tuberosity, is the true insertion of the quadriceps and the patella is thus a sesamoid bone (the largest in the body). This arrangement constitutes the *extensor mechanism*.

• The posterior surface of the patella is smooth and covered with articular cartilage. It is divided into a large lateral facet and a smaller medial facet for articulation with the femoral condyles.

Bones of the foot

See 'The Foot Bones' Chapter 55, p. 129.

Clinical notes

• **Fracture of the patella**: a violent contraction of quadriceps can cause a transverse fracture of the patella, rupture of the ligamentum patellae or avulsion of the tibial tuberosity.

• **Dislocation of the patella**: because of the obliquity of the femur, the line of pull of quadriceps is upwards and laterally, whereas the ligamentum patellae is vertical. There is thus a tendency for the patella to be displaced laterally, so that a strong contraction of the muscle can cause a dislocation, the patella coming to lie lateral to the knee joint.

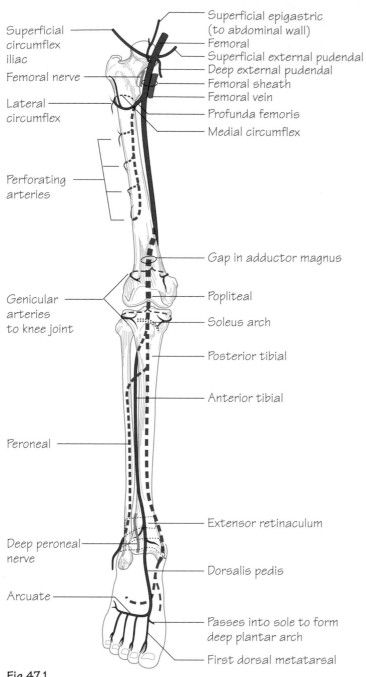

Superficial circumflex iliac

Femoral nerve

Lateral circumflex

Perforating arteries

Genicular arteries to knee joint

Peroneal

Deep peroneal nerve

Arcuate

Superficial epigastric (to abdominal wall)

Femoral

Superficial external pudendal

Deep external pudendal

Femoral sheath

Femoral vein

Profunda femoris

Medial circumflex

Gap in adductor magnus

Popliteal

Soleus arch

Posterior tibial

Anterior tibial

Extensor retinaculum

Dorsalis pedis

Passes into sole to form deep plantar arch

First dorsal metatarsal

Fig.47.1
The course and major branches of the femoral artery

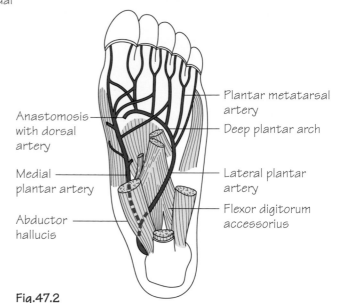

Anastomosis with dorsal artery

Medial plantar artery

Abductor hallucis

Plantar metatarsal artery

Deep plantar arch

Lateral plantar artery

Flexor digitorum accessorius

Fig.47.2
The medial and lateral plantar arteries

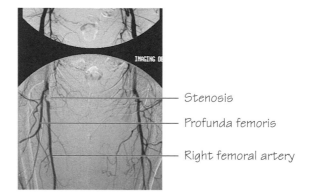

Stenosis

Profunda femoris

Right femoral artery

Fig.47.3
An angiogram of the lower limbs showing stenosis of the femoral artery on the right side.
(The profunda is often known as the deep femoral and the continuation of the femoral artery as the superficial femoral)

The femoral artery

- **Course**: the *femoral artery* commences as a continuation of the *external iliac artery* behind the inguinal ligament at the mid-inguinal point (halfway between the anterior superior iliac spine and the symphysis pubis). In the groin, the femoral vein lies immediately medial to the artery and both are enclosed in the femoral sheath. In contrast, the femoral nerve lies immediately lateral to the femoral sheath. The femoral artery descends the thigh to pass under sartorius and then through the *adductor (Hunter's) canal* to become the *popliteal artery*.
- **Branches**:
  - *Branches in the upper part of the femoral triangle*: four branches are given off which supply the superficial tissues of the lower abdominal wall and perineum (see Fig. 47.1).
  - *Profunda femoris*: arises from the lateral side of the femoral artery 4 cm below the inguinal ligament. Near its origin it gives rise to *medial* and *lateral circumflex femoral branches*. These contribute to the *trochanteric* and *cruciate anastomoses* (see below). The profunda descends deep to adductor longus in the medial compartment of the thigh and gives rise to four *perforating* branches. These circle the femur posteriorly, perforating and supplying all muscles in their path. The profunda and perforating branches ultimately anastomose with the genicular branches of the popliteal artery.

The trochanteric anastomosis

This arterial anastomosis is formed by branches from the *medial* and *lateral circumflex femoral*, the *superior gluteal* and, usually, the *inferior gluteal arteries*. It lies close to the trochanteric fossa and provides branches that ascend the femoral neck beneath the retinacular fibres of the capsule to supply the femoral head.

The cruciate anastomosis

This anastomosis constitutes a collateral supply. It is formed by: the *transverse* branches of the *medial* and *lateral circumflex femoral arteries*, the *descending* branch of the *inferior gluteal artery* and the *ascending* branch of the *1st perforating* branch of the profunda.

The popliteal artery

- **Course**: the *femoral artery* continues as the *popliteal artery* as it passes through the hiatus in adductor magnus to enter the popliteal fossa. From above, it descends on the posterior surface of the femur, the capsule of the knee joint and the fascia overlying popliteus to pass under the fibrous arch of soleus, where it bifurcates into *anterior* and *posterior tibial arteries*. The popliteal artery is the deepest structure in the popliteal fossa, rendering its pulsations difficult to feel. The *popliteal vein* crosses the artery superficially and the *tibial nerve* crosses from lateral to medial over the vein.
- **Branches**: *muscular*, *sural* and five *genicular* arteries are given off. The last forms a rich anastomosis around the knee.

The anterior tibial artery

- **Course**: the *anterior tibial artery* passes anteriorly from its origin, accompanied by its venae comitantes, over the upper border of the interosseous membrane and then descends over the anterior surface of the membrane giving off muscular branches to the extensor compartment of the leg. The artery crosses the front of the ankle joint midway between the malleoli where it becomes the *dorsalis pedis artery*. Tibialis anterior and extensor digitorum longus flank the artery throughout its course on its medial and lateral sides, respectively. Extensor hallucis longus commences on the lateral side, but crosses the artery to lie medial by the end of its course. The dorsalis pedis artery passes on the dorsum of the foot to the level of the base of the metatarsals and then dives between the two heads of the first dorsal interosseous muscle to gain access to the sole and complete the deep plantar arch. Prior to passing to the sole, it gives off the *1st dorsal metatarsal* branch and, via an *arcuate* branch, the three remaining *dorsal metatarsal* branches (Fig. 47.1).
- **Branches of the anterior tibial artery include**: *muscular* and *malleolar* branches.

The posterior tibial artery

- **Course**: the *posterior tibial artery* arises as a terminal branch of the popliteal artery. It is accompanied by its venae comitantes and supplies the flexor compartment of the leg. Approximately midway down the calf the tibial nerve crosses behind the artery from medial to lateral. The artery ultimately passes behind the medial malleolus to divide into *medial* and *lateral plantar arteries* under the flexor retinaculum. The latter branches gain access to the sole deep to abductor hallucis. Posterior to the medial malleolus, the structures which can be identified – from front to back – are: tibialis posterior, flexor digitorum longus, posterior tibial artery and venae comitantes, the tibial nerve and flexor hallucis longus.
- **Branches**:
  - *Peroneal artery*: this artery usually arises from the posterior tibial artery approximately 2.5 cm along its length. It courses between tibialis posterior and flexor hallucis longus and supplies the peroneal (lateral) compartment of the leg. It ends by dividing into a *perforating branch* that pierces the interosseous membrane and a *lateral calcaneal branch*.
  - *Other branches*: the *posterior tibial artery* gives rise to *nutrient* and *muscular* branches throughout its course.
  - *Lateral plantar artery*: passes between flexor accessorius and flexor digitorum brevis to the lateral aspect of the sole where it divides into *superficial* and *deep* branches. The deep branch runs between the 3rd and 4th muscle layers of the sole to continue as the *deep plantar arch* which is completed by the termination of the dorsalis pedis artery. The arch gives rise to *plantar metatarsal* branches which supply the toes (Fig. 47.2).
  - *Medial plantar artery*: runs on the medial aspect of the sole and sends branches which join with the plantar metatarsal branches of the lateral plantar artery to supply the toes.

Clinical notes

- **Peripheral vascular disease (Fig. 47.3)**: atheroma causes narrowing of the peripheral arteries with a consequent reduction in flow. Whilst flow may be adequate for tissue perfusion at rest, exercise causes pain due to ischaemia (*intermittent claudication*). When symptoms are intolerable, pain is present at rest or ischaemic ulceration occurs; arterial reconstruction is required. Reconstruction is performed using either the patient's own saphenous vein or a synthetic graft (Dacron or PTFE) to bypass the occlusion. Disease, limited in extent, may be suitable for interventional procedures, such as percutaneous transluminal angioplasty (PTA) or stent insertion.

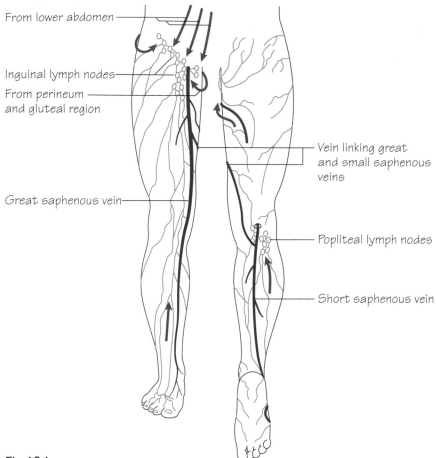

Fig.48.1
The superficial veins and lymphatics of the lower limb.
The arrows indicate the direction of lymph flow

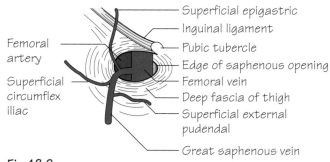

Fig.48.2
The termination of the great saphenous vein

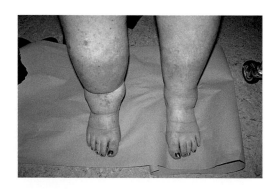

Fig.48.3
Lymphoedema of the lower limb

The superficial veins of the lower limb
(Fig. 48.1)

The superficial system comprises the *great* and *small saphenous veins*. These are of utmost clinical importance as they are predisposed towards becoming varicose and, consequently, often require surgery. They are also the commonly used conduits for coronary artery surgery.

- The *great saphenous vein* arises from the medial end of the dorsal venous network on the foot. It passes anterior to the medial malleolus, along the anteromedial aspect of the calf (with the saphenous nerve), migrates posteriorly to a hand's breadth behind the patella at the knee and then courses forwards to ascend the anteromedial thigh. It pierces the cribriform fascia to drain into the femoral vein at the saphenous opening. The terminal part of the great saphenous vein usually receives superficial tributaries from the external genitalia and the lower abdominal wall (Fig. 48.2). At surgery, these help to distinguish the saphenous from the femoral vein as the only tributary draining into the latter is the saphenous vein. *Anteromedial* and *posterolateral femoral* (*lateral accessory*) tributaries, from the medial and lateral aspects of the thigh, also sometimes drain into the great saphenous vein below the saphenous opening.

The great saphenous vein is connected to the deep venous system at multiple levels by *perforating veins*. These usually occur above and below the medial malleolus, in the 'gaiter area', in the mid-calf region, below the knee and one long connection in the lower thigh. The valves in the perforators are directed inwards, so that blood flows from superficial to deep systems, from where it can be pumped upwards, assisted by the muscular contractions of the calf muscles. The deep system is consequently at higher pressure than the superficial and, thus, should the valves in the perforators become incompetent, the increased pressure is transmitted to the superficial system and these veins become varicose.

- The *small saphenous vein* arises from the lateral end of the dorsal venous network on the foot. It passes behind the lateral malleolus and over the back of the calf to pierce the deep fascia in an inconstant position to drain into the popliteal vein.

The deep veins of the lower limb

The deep veins of the calf are the *venae comitantes* of the *anterior* and *posterior tibial arteries,* which go on to become the *popliteal* and *femoral veins*. The deep veins form an extensive network within the posterior compartment of the calf – the *soleal plexus* – from which blood is assisted upwards against gravitational forces by muscular contraction during exercise (the 'muscle pump').

The lymphatics of the lower limb
(Fig. 48.1)

The lymph nodes of the groin are arranged into *superficial* and *deep* groups. The superficial inguinal group lies in the superficial fascia and is arranged into two chains:

- *Longitudinal chain*: the lymph nodes in this chain lie along the terminal portion of the great saphenous vein. They receive lymph from the majority of the superficial tissues of the lower limb.
- *Horizontal chain*: the lymph nodes in this chain lie parallel to the inguinal ligament. They receive lymph from the superficial tissues of the lower trunk below the level of the umbilicus, the buttock, the external genitalia and the lower half of the anal canal. The superficial nodes drain into the deep nodes through the saphenous opening in the deep fascia.

The deep inguinal nodes are situated medial to the femoral vein. They are usually three in number. These nodes receive lymph from all of the tissues deep to the fascia lata of the lower limb. In addition, they also receive lymph from the skin and superficial tissues of the heel and lateral aspect of the foot by way of the popliteal nodes. The deep nodes convey lymph to the external iliac and thence to the para-aortic nodes.

Clinical notes

- **Varicose veins**: these are classified as follows:
- **Primary**: there is a deficiency of collagen and elastic tissue in the walls of the veins; the veins become dilated and the valves become incompetent in both the superficial veins and perforators.
- **Secondary**: this is the result of obstruction to the deep venous drainage, for example, by pressure from the fetal head on the pelvic veins in pregnancy or by deep venous thrombosis.
- **Deep venous thrombosis**: interference with the action of the muscle pump in returning blood to the heart leads to pooling of blood in the lower limb and the possibility of *deep venous thrombosis*. This can occur, for instance, in prolonged immobilisation in bed or in the cramped conditions of long air journeys. This is a potentially life-threatening condition owing to the possibility of parts of the clot breaking off and travelling via the right side of the heart to the lungs, causing *pulmonary embolism*.
- **Lymphoedema**: obstruction of the lymphatics results in lymphoedema (Fig. 48.3). This can be congenital, as a result of aberrant lymphatic formation, or acquired, such as post-radiotherapy or following certain infections. In developing countries, infection with *Filaria bancrofti* is a significant cause of lymphoedema that can progress to massive proportions, requiring limb reduction or even amputation.

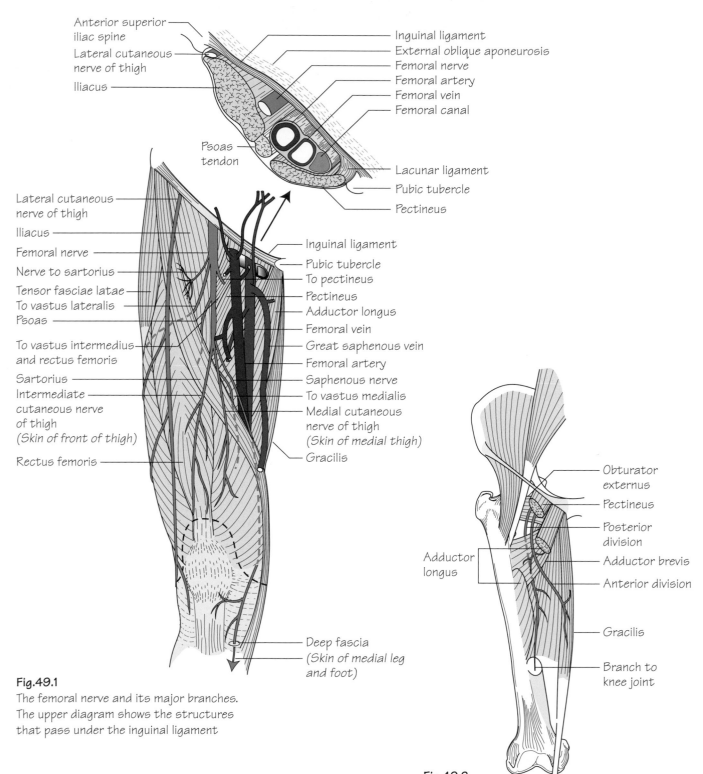

Fig.49.1
The femoral nerve and its major branches.
The upper diagram shows the structures
that pass under the inguinal ligament

Fig.49.2
The anterior and posterior divisions
of the obturator nerve

The lumbar plexus (T12–L4) (see Fig. 25.1)

See Chapter 25.

- **Origins**: from the anterior primary rami of T12–L4.
- **Course**: the majority of the branches of the plexus passes through the substance of psoas major and emerge at its lateral border except for the genitofemoral and obturator nerves.
- **Branches**:
  - *Intra-abdominal branches*: these are described in Chapter 25.
  - *Femoral nerve (L2, L3 and L4)*: see below.
  - *Obturator nerve (L2, L3 and L4)*: see below.
  - *Lateral cutaneous nerve of the thigh (L2 and L3)*: crosses the iliac fossa over iliacus and passes under the lateral part of the inguinal ligament to enter the superficial tissue of the lateral thigh which it innervates with sensory fibres.

The femoral nerve (L2, L3 and L4)
(Fig. 49.1)

- **Origins**: the *posterior divisions* of the anterior primary rami of L2, L3 and L4.
- **Course**: the femoral nerve traverses psoas to emerge at its lateral border. It descends through the iliac fossa to pass under the inguinal ligament. At this point it lies on iliacus, which it supplies, and is situated immediately lateral to the femoral sheath. It branches within the femoral triangle only a short distance (5 cm) beyond the inguinal ligament. The *lateral circumflex femoral artery* passes through these branches to divide them into *superficial* and *deep* divisions:
  - *Superficial division*: consists of *medial* and *intermediate cutaneous branches*, which supply the skin over the anterior and medial aspects of the thigh, and two muscular branches. The latter supply sartorius and pectineus.
  - *Deep division*: consists of four muscular branches which supply the components of quadriceps femoris, and the joints over which they pass, and one cutaneous nerve – the *saphenous nerve*. The latter nerve is the only branch to extend beyond the knee. It pierces the deep fascia overlying the adductor canal and descends through the leg, accompanied by the great saphenous vein, to supply the skin over the medial aspect of the leg and foot.

The obturator nerve (L2, L3 and L4)
(Fig. 49.2)

- **Origins**: the *anterior divisions* of the anterior primary rami of L2, L3 and L4.
- **Course**: obturator nerve emerges at the medial border of psoas (cf. other nerves which traverse psoas to emerge at the lateral border). It passes over the pelvic brim to pass through the upper aspect of the *obturator foramen* with other obturator vessels. In the obturator notch, it divides into *anterior* and *posterior* divisions which pass in front of and behind adductor brevis to supply the muscles of the adductor compartment:
  - *Anterior division*: gives rise to an articular branch to the hip joint as well as muscular branches to adductor longus, brevis and gracilis. It terminates by supplying the skin of the medial aspect of the thigh.
  - *Posterior division*: supplies muscular branches to obturator externus, adductor brevis and magnus, as well as an articular branch to the knee.

Clinical notes

- **Meralgia paraesthetica**: obese patients sometimes describe paraesthesiae over the lateral thigh. This is termed *meralgia paraesthetica* and results from compression of the lateral cutaneous nerve of the thigh as it passes under (or sometimes through) the inguinal ligament.
- **Referred pain**: the obturator nerve supplies both the hip and the knee joints, as does the femoral nerve. For this reason, pain from disease of the hip joint may sometimes be referred to the knee and vice versa.

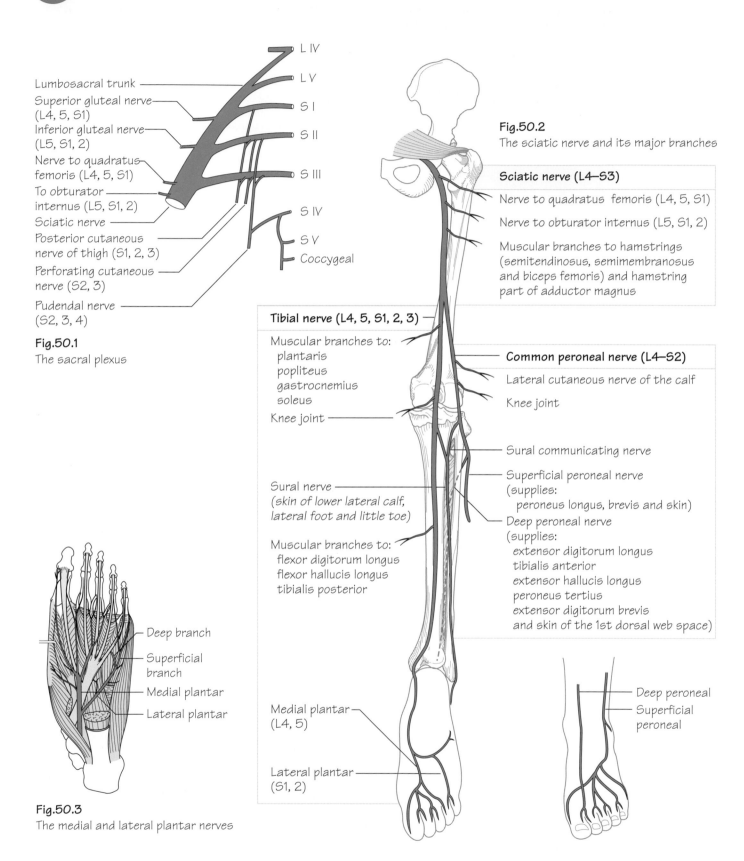

Lumbosacral trunk

Superior gluteal nerve
(L4, 5, S1)

Inferior gluteal nerve
(L5, S1, 2)

Nerve to quadratus
femoris (L4, 5, S1)

To obturator
internus (L5, S1, 2)

Sciatic nerve

Posterior cutaneous
nerve of thigh (S1, 2, 3)

Perforating cutaneous
nerve (S2, 3)

Pudendal nerve
(S2, 3, 4)

L IV
L V
S I
S II
S III
S IV
S V
Coccygeal

Fig.50.1
The sacral plexus

Fig.50.2
The sciatic nerve and its major branches

Sciatic nerve (L4–S3)

Nerve to quadratus femoris (L4, 5, S1)

Nerve to obturator internus (L5, S1, 2)

Muscular branches to hamstrings
(semitendinosus, semimembranosus
and biceps femoris) and hamstring
part of adductor magnus

Tibial nerve (L4, 5, S1, 2, 3)

Muscular branches to:
plantaris
popliteus
gastrocnemius
soleus

Knee joint

Sural nerve
(skin of lower lateral calf,
lateral foot and little toe)

Muscular branches to:
flexor digitorum longus
flexor hallucis longus
tibialis posterior

Common peroneal nerve (L4–S2)

Lateral cutaneous nerve of the calf

Knee joint

Sural communicating nerve

Superficial peroneal nerve
(supplies:
peroneus longus, brevis and skin)

Deep peroneal nerve
(supplies:
extensor digitorum longus
tibialis anterior
extensor hallucis longus
peroneus tertius
extensor digitorum brevis
and skin of the 1st dorsal web space)

Medial plantar
(L4, 5)

Lateral plantar
(S1, 2)

Deep peroneal
Superficial
peroneal

Deep branch

Superficial
branch

Medial plantar

Lateral plantar

Fig.50.3
The medial and lateral plantar nerves

The sacral plexus (L4–S4) (Fig. 50.1)

- **Origins**: from the anterior primary rami of L4–S4.
- **Course**: the sacral nerves emerge through the anterior sacral foramina. The nerves unite and are joined by the lumbosacral trunk (L4,5), anterior to piriformis.
- **Branches**: the branches of the sacral plexus include:
  - **The superior gluteal nerve (L4, L5 and S1)**: arises from the roots of the sciatic nerve and passes through the greater sciatic foramen above the upper border of piriformis. In the gluteal region, it runs below the middle gluteal line between gluteus medius and minimus (both of which it supplies) before terminating in the substance of tensor fasciae latae.
  - **The inferior gluteal nerve (L5, S1 and S2)**: arises from the roots of the sciatic nerve and passes through the greater sciatic foramen below piriformis. In the gluteal region, it penetrates and supplies gluteus maximus.
  - **The posterior cutaneous nerve of the thigh (S1, S2 and S3)**: passes through the greater sciatic foramen below piriformis. Its branches supply the skin of the scrotum, buttock and back of the thigh up to the knee.
  - **The perforating cutaneous nerve (S2 and S3)**: perforates gluteus maximus to supply the skin of the buttock.
  - **The pudendal nerve (S2, S3 and S4)**: passes briefly into the gluteal region by passing out of the greater sciatic foramen below piriformis over the sacrospinous ligament and passes back into the pelvis through the lesser sciatic foramen. It runs forwards in the pudendal (Alcock's) canal and gives off its inferior rectal branch in the ischio-rectal fossa. It continues its course to the perineum and divides into dorsal and perineal branches that pass deep and superficial to the urogenital diaphragm, respectively.
  - **The sciatic nerve**: see below.

The sciatic nerve (L4–S3) (Fig. 50.2)

- **Origins**: the anterior primary rami of L4, L5, S1, S2 and S3.
- **Course**: the sciatic nerve passes through the greater sciatic foramen below piriformis under the cover of gluteus maximus. In the gluteal region, it passes over the superior gemellus, obturator internus and inferior gemellus and then over quadratus femoris and adductor magnus in the thigh as it descends in the midline. The sciatic nerve divides into its terminal branches, the tibial and common peroneal nerves, usually just below the mid-thigh, although a higher division is not uncommon.
- **Branches**:
  - **Muscular branches**: to supply the hamstrings and the ischial part of adductor magnus.
  - **Nerve to obturator internus (L5,S1,2)**: supplies obturator internus and the superior gemellus.
  - **Nerve to quadratus femoris (L4,5,S1)**: supplies quadratus femoris and the inferior gemellus.
  - **Tibial nerve**: see below.
  - **Common peroneal nerve**: see below.

The tibial nerve (L4–S3) (Fig. 50.2)

- **Origins**: it is a terminal branch of the sciatic nerve.
- **Course**: it traverses the popliteal fossa over the popliteal vein and artery from the lateral to medial side. It leaves the popliteal fossa by passing under the fibrous arch of soleus and, in the leg, descends with the posterior tibial artery under the cover of this muscle. The nerve crosses the posterior tibial artery from medial to lateral in the mid-calf and, together with the artery, passes behind the medial malleolus and then under the flexor retinaculum where it divides into its terminal branches, the *medial* and *lateral plantar* nerves.

- **Main branches**:
  - **Genicular branches**: to the knee joint.
  - **Muscular branches**: to plantaris, soleus, gastrocnemius and the deep muscles at the back of the leg.
  - **Sural nerve**: arises in the popliteal fossa and is joined by the sural communicating branch of the common peroneal nerve. It pierces the deep fascia in the calf and descends subcutaneously with the small saphenous vein. It passes behind the lateral malleolus and under the flexor retinaculum to divide into its cutaneous terminal branches which supply the skin of the lower lateral calf, foot and little toe.
  - **Medial plantar nerve (L4 and L5) (Fig. 50.3)**: runs with the medial plantar artery between abductor hallucis and flexor digitorum brevis. It sends four motor branches and a cutaneous supply to the medial 3.5 digits.
  - **Lateral plantar nerve (S1 and S2) (Fig. 50.3)**: runs with the lateral plantar artery to the base of the 5th metatarsal where it divides into superficial and deep branches. These collectively supply the skin of the lateral 1.5 digits and the remaining muscles of the sole.

The common peroneal nerve (L4–S2) (Fig. 50.2)

- **Origin**: a terminal branch of the sciatic nerve.
- **Course**: it passes along the medial border of the biceps femoris tendon along the superolateral margin of the popliteal fossa. The nerve then winds around the neck of the fibula (see Fig. 46.4) and, in the substance of peroneus longus, divides into its terminal branches, the *superficial* and *deep peroneal nerves*.
- **Branches**:
  - **Genicular branches** to the knee joint.
  - **Lateral cutaneous nerve of the calf.**
  - **A sural communicating branch.**
  - **Superficial peroneal nerve (L5, S1 and S2)**: this branch runs in and supplies the muscles of the lateral (peroneal) compartment of the leg. In addition, it supplies the skin over the lateral lower two-thirds of the leg and the whole of the dorsum of the foot, except for the area between the 1st and 2nd toes, which is supplied by the deep peroneal nerve.
  - **Deep peroneal nerve (L4,5,S1,2)**: runs with the anterior tibial vessels over the interosseous membrane into the anterior compartment of the leg and then over the ankle to the dorsum of the foot. It supplies all of the muscles of the anterior compartment as well as providing a cutaneous supply to the area between the 1st and 2nd toes.

Clinical notes

- **Foot drop**: the common peroneal nerve is exposed to injury as it winds around the neck of the fibula, for example, by fracture of the fibular neck. The resultant paralysis of the dorsiflexor muscles leads to *foot drop*. The patient walks with a high-stepping gait so as to lift the dropped foot clear of the ground. The toes of the shoes are often scuffed due to dragging of the foot along the ground.

51 The hip joint and gluteal region

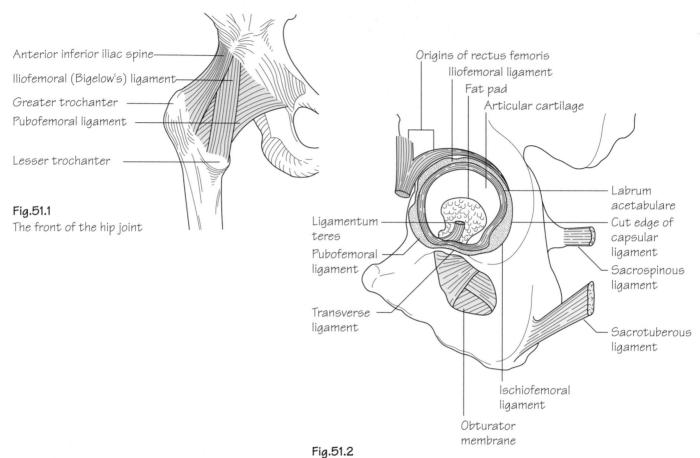

Anterior inferior iliac spine

Iliofemoral (Bigelow's) ligament

Greater trochanter

Pubofemoral ligament

Lesser trochanter

Fig.51.1
The front of the hip joint

Origins of rectus femoris
Iliofemoral ligament
Fat pad
Articular cartilage

Ligamentum teres

Pubofemoral ligament

Transverse ligament

Labrum acetabulare

Cut edge of capsular ligament

Sacrospinous ligament

Sacrotuberous ligament

Ischiofemoral ligament

Obturator membrane

Fig.51.2
The structures around the acetabulum

The hip joint (Figs. 51.1, 51.2, and 51.3)

• **Type**: the hip is a *synovial ball and socket joint*. The articulation is between the rounded femoral head and the acetabulum which, like the shoulder, is deepened at its margins by a fibrocartilaginous rim – the *labrum acetabulare*. The central and inferior parts of the acetabulum are devoid of articulating surface. This region is termed the *acetabular notch* from which the *ligamentum teres* passes to the *fovea* on the femoral head. The inferior margin below the acetabular notch is completed by the *transverse acetabular ligament*.

• **Capsule**: the capsule of the hip joint is attached above to the acetabular margin, including the transverse acetabular ligament. The capsule attaches to the femur anteriorly at the *intertrochanteric line* and to the bases of the trochanters. Posteriorly, the capsule attaches to the femur at a higher level – approximately 1 cm above the *trochanteric crest*. The capsular fibres are reflected from the lower attachment upwards on the femoral neck as *retinacula*. These fibres are of extreme importance as they carry with them a blood supply to the femoral head.

• **Stability**: the stability of the hip is dependent predominantly on bony factors. Ligamentous stability is provided by three ligaments:

  1 *Iliofemoral* ligament (*Bigelow's ligament*): is inverted Y-shaped and strong. It arises from the *anterior inferior iliac spine* and inserts at either end of the *intertrochanteric line*. This ligament restricts hyperextension at the hip.

  2 *Pubofemoral* ligament: arises from the iliopubic junction and passes to the capsule over the intertrochanteric line where it attaches.

  3 *Ischiofemoral* ligament: fibres arise from the ischium and some encircle laterally to attach to the base of the greater trochanter. The majority of the fibres, however, spiral and blend with the capsule around the neck of the femur – the *zona orbicularis*.

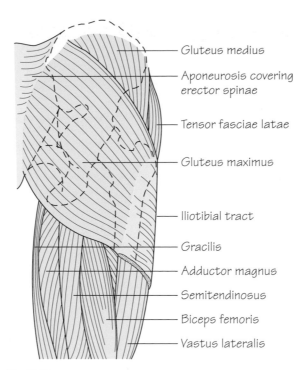

Fig.51.3
The superficial muscles of the gluteal region

- **Synovium**: the synovial membrane lines the capsule of the hip joint and is reflected back along the femoral neck. It invests the ligamentum teres as a sleeve and attaches to the articular margins. A *psoas bursa* occurs in 10% of the population. This is an outpouching of synovial membrane through a defect in the anterior capsular wall under the psoas tendon.
- **Blood supply (Fig. 51.6)**: the femoral head derives its blood supply from three main sources:
  1 Vessels which pass along the neck with the capsular retinacula and enter the head through large foramina at the base of the head. These are derived from branches of the circumflex femoral arteries via the cruciate and trochanteric anastomoses. This is the most important supply in the adult.
  2 Vessels in the ligamentum teres which enter the head through small foramina in the fovea. These are derived from branches of the obturator artery.
  3 Through the diaphysis via nutrient femoral vessels.
- **Nerve supply**: is from branches of the femoral, sciatic and obturator nerves.

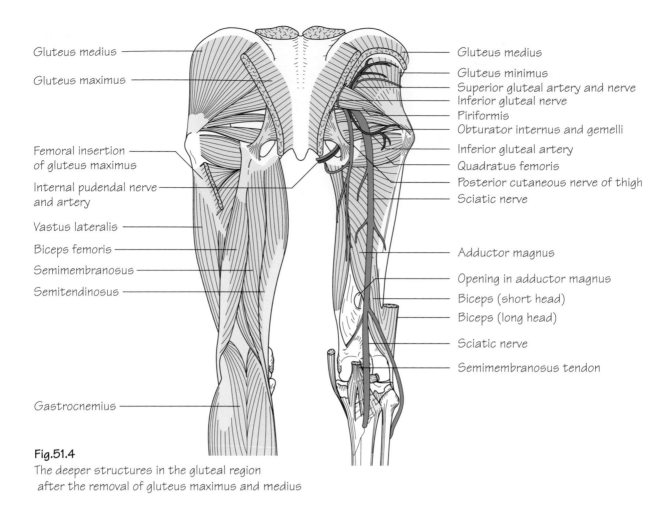

Gluteus medius

Gluteus maximus

Femoral insertion
of gluteus maximus

Internal pudendal nerve
and artery

Vastus lateralis

Biceps femoris

Semimembranosus

Semitendinosus

Gastrocnemius

Gluteus medius
Gluteus minimus
Superior gluteal artery and nerve
Inferior gluteal nerve
Piriformis
Obturator internus and gemelli
Inferior gluteal artery
Quadratus femoris
Posterior cutaneous nerve of thigh
Sciatic nerve

Adductor magnus

Opening in adductor magnus

Biceps (short head)

Biceps (long head)

Sciatic nerve

Semimembranosus tendon

Fig.51.4
The deeper structures in the gluteal region
after the removal of gluteus maximus and medius

Hip movements

A wide range of movement is possible at the hip because of its ball and socket articulation.

- **Flexion (0–120°)**: iliacus and psoas predominantly. Rectus femoris, sartorius and pectineus to a lesser degree.
- **Extension (0–20°)**: gluteus maximus and the hamstrings.
- **Adduction (0–30°)**: adductor magnus, longus and brevis predominantly. Gracilis and pectineus to a lesser degree.

- **Abduction (0–45°)**: gluteus medius, gluteus minimus and tensor fasciae latae.
- **Lateral rotation (0–45°)**: piriformis, obturators, the gemelli, quadratus femoris and gluteus maximus.
- **Medial rotation (0–45°)**: tensor fasciae latae, gluteus medius and gluteus minimus.
- **Circumduction**: this is a combination of all movements utilising all muscle groups mentioned.

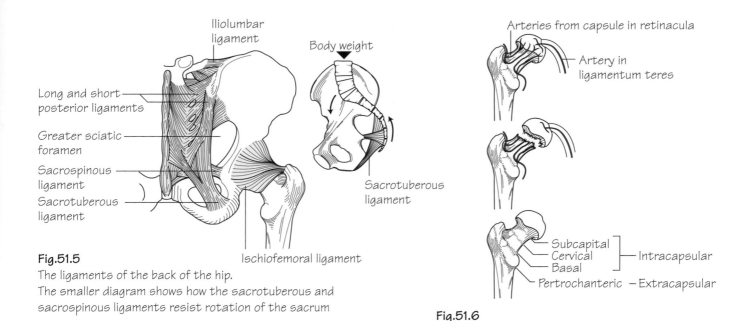

Fig.51.5
The ligaments of the back of the hip.
The smaller diagram shows how the sacrotuberous and
sacrospinous ligaments resist rotation of the sacrum

Fig.51.6
The terminology of fractures of the neck of the femur.
Fractures near the head can cause avascular
necrosis because of the disruption of the arterial
supply to the head

The gluteal region (Figs. 51.5 and 51.6)

The gluteal region is limited above by the iliac crest and below by the transverse skin crease – the *gluteal fold*. The fold occurs as the overlying skin is bound to the underlying deep fascia and not, as is often thought, by the contour of gluteus maximus. The greater and lesser sciatic foramina are formed by the pelvis and the sacrotuberous and sacrospinous ligaments (Fig. 51.5). Through these, structures pass from the pelvis to the gluteal region.

Contents of the gluteal region (Fig. 51.4)

- **Muscles**: of the gluteal region include the *gluteus maximus, gluteus medius, gluteus minimus, tensor fasciae latae, piriformis, gemellus superior, gemellus inferior, obturator internus* and *quadratus femoris* (see Muscle index, p. xx).
- **Nerves**: of the gluteal region include the *sciatic nerve* (L4, L5, S1, S2 and S3), *posterior cutaneous nerve of the thigh, superior* (L4, L5, S1 and S2) and *inferior* (L5,S1 and S2) *gluteal nerves, nerve to quadratus femoris* (L4, L5 and S1) and the *pudendal nerve* (S2, S3 and S4).
- **Arteries**: of the gluteal region include the *superior* and *inferior gluteal arteries*. These anastomose with the medial and lateral femoral circumflex arteries, and the first perforating branch of the profunda, to form the trochanteric and cruciate anastomoses, respectively.

Clinical notes

- **Fractured neck of femur** (Fig. 51.6): femoral neck fractures are common following falls amongst the elderly osteoporotic population. Fractures in this region present a considerable risk of *avascular necrosis* if the fracture line is intracapsular, as the retinacula, which carry the main arterial supply, are torn. In contrast, extracapsular femoral neck fractures present no risk of avascular necrosis. If the fracture components are not impacted, the usual clinical presentation is that of shortening and external rotation of the affected limb. This occurs as the adductors, hamstrings and rectus femoris pull upwards on the distal fragment, whilst piriformis, the gemelli, obturators, gluteus maximus and gravity produce lateral rotation.
- **Trendelenberg's sign**: the abductor muscles (gluteus medius and minimus and tensor fasciae latae) not only abduct the leg but, by acting from insertion to origin, can tilt the pelvis towards the same side or just support it when the opposite leg is lifted from the ground. The lever on which the muscles act is the head and neck of the femur. If there is any weakness of the muscles or distortion of the neck of the femur, for example as a result of disease of the femoral head or an old fracture, when the patient stands on one leg the opposite side of the pelvis will drop (*Trendelenberg's sign*). The patient walks with a characteristic waddling gait.
- **Intramuscular injections**: the gluteal region is a common site for intramuscular injections. To avoid possible damage to the sciatic nerve, the safest site for such injections is the upper and outer quadrant of the gluteal region.

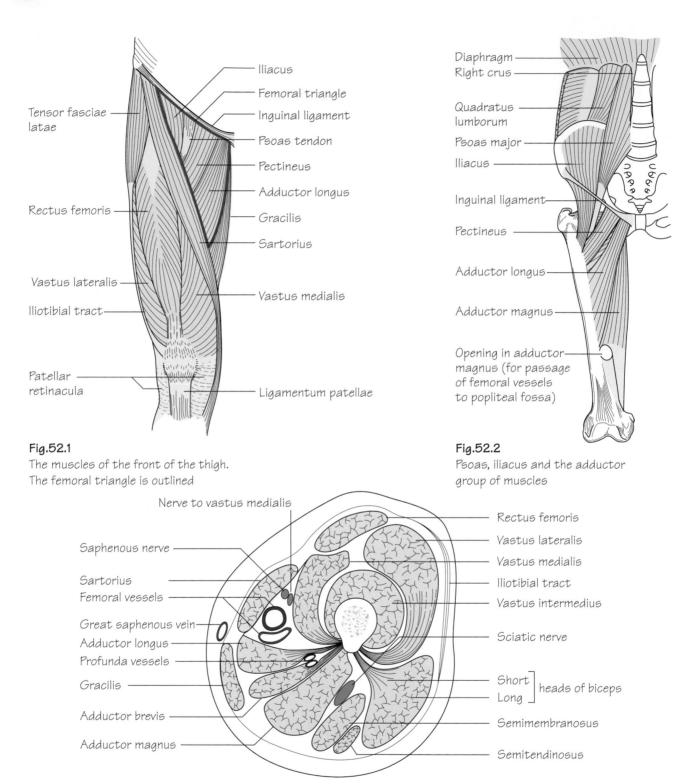

Fig.52.1
The muscles of the front of the thigh.
The femoral triangle is outlined

Labels (Fig.52.1):
Tensor fasciae latae
Iliacus
Femoral triangle
Inguinal ligament
Psoas tendon
Pectineus
Adductor longus
Gracilis
Sartorius
Rectus femoris
Vastus lateralis
Iliotibial tract
Vastus medialis
Patellar retinacula
Ligamentum patellae

Fig.52.2
Psoas, iliacus and the adductor
group of muscles

Labels (Fig.52.2):
Diaphragm
Right crus
Quadratus lumborum
Psoas major
Iliacus
Inguinal ligament
Pectineus
Adductor longus
Adductor magnus
Opening in adductor magnus (for passage of femoral vessels to popliteal fossa)

Fig.52.3
A section through the thigh to show the adductor (subsartorial) canal

Labels (Fig.52.3):
Nerve to vastus medialis
Saphenous nerve
Sartorius
Femoral vessels
Great saphenous vein
Adductor longus
Profunda vessels
Gracilis
Adductor brevis
Adductor magnus
Rectus femoris
Vastus lateralis
Vastus medialis
Iliotibial tract
Vastus intermedius
Sciatic nerve
Short / Long] heads of biceps
Semimembranosus
Semitendinosus

The thigh is divided into flexor, extensor and adductor compartments. The membranous superficial fascia of the abdominal wall fuses to the *fascia lata*, the deep fascia of the lower limb, at the skin crease of the hip joint just below the inguinal ligament.

The deep fascia of the thigh (fascia lata)

This layer of strong fascia covers the thigh. It is attached above to the inguinal ligament and bony margins of the pelvis and below to the tibial condyles, head of the fibula and patella. Three fascial septa pass from the deep surface of the fascia lata to insert onto the linea aspera of the femur and consequently divide the thigh into three compartments.

On the lateral side, the fascia lata is condensed to form the *iliotibial tract* (Fig. 52.4). The tract is attached above to the iliac crest and receives the insertions of tensor fasciae latae and three-quarters of gluteus maximus. These muscles are also enveloped in deep fascia. The iliotibial tract inserts into the lateral condyle of the tibia.

The *saphenous opening* is a gap in the deep fascia which is filled with loose connective tissue – the *cribriform fascia*. The lateral border of the opening, the *falciform margin*, curves in front of the femoral vessels, whereas, on the medial side, it curves behind to attach to the iliopectineal line (Fig. 48.2). The great saphenous vein pierces the cribriform fascia to drain into the femoral vein. Superficial branches of the femoral artery and lymphatics are also transmitted through the saphenous opening.

The superficial fascia of the thigh

Contents of the subcutaneous tissue include:
- **Nerves**: the femoral branch of the genitofemoral nerve (p. xx), the medial, intermediate (branches of the femoral nerve, p. xx) and lateral femoral cutaneous nerves (L2,3, p. xx) and branches of the obturator nerve (p. xx) supply the skin of the anterior thigh. The back of the thigh receives its sensory supply from the posterior cutaneous nerve of the thigh.
- **Superficial arteries**: these include the four superficial branches of the femoral artery: the superficial circumflex iliac artery, superficial epigastric artery, superficial external pudendal artery and the deep external pudendal artery.
- **Superficial veins and lymphatics**: venous tributaries of the anterior thigh drain into the great saphenous vein, whilst some in the lower posterior thigh drain into the popliteal vein. The great saphenous vein is also accompanied by large lymphatics which pass to the superficial inguinal nodes and, from there, through the cribriform fascia to the deep inguinal nodes.

The femoral triangle (Figs. 49.1 and 52.1)

The boundaries of the femoral triangle are: the inguinal ligament above, the *medial* border of sartorius and the *medial* border of adductor longus.
- **The floor consists of** the adductor longus, pectineus, psoas tendon and iliacus (see Muscle index, p. xx).
- **The roof consists of** the fascia lata. The saphenous opening is in the upper part of the triangle.
- **The contents include** (from lateral to medial) the femoral nerve, artery, vein and their branches and tributaries. The *femoral canal* is situated medial to the femoral vein. Transversalis fascia and psoas fascia fuse and evaginate to form the *femoral sheath* below the inguinal ligament. The sheath encloses the femoral artery, vein and canal but the femoral nerve lies outside on its lateral aspect (see Fig. 49.1).

The contents of the anterior compartment of the thigh
(Figs. 52.1–52.3)
- **Muscles**: these constitute the hip flexors and knee extensors, i.e. *sartorius, iliacus, psoas, pectineus* and *quadriceps femoris* (see Muscle index, p. xx).
- **Arteries**: the femoral artery and its branches (p. xx).
- **Veins**: the femoral vein is a continuation of the popliteal vein as it passes through the hiatus in adductor magnus. It receives its main tributary – the *great saphenous* vein – through the saphenous opening.
- **Lymphatics**: from the anterior compartment, pass to the deep inguinal lymph nodes which lie along the terminal part of the femoral vein.
- **Nerves**: the femoral nerve (L2, L3 and L4, p. xx) divides a short distance below the inguinal ligament into superficial and deep divisions. Only the saphenous branch passes beyond the knee.

The contents of the medial compartment of the thigh (Figs. 52.2 and 52.3)
- **Muscles**: these comprise the hip adductors: *gracilis, adductor longus, adductor brevis, adductor magnus* and *obturator externus* (a lateral rotator of the thigh at the hip) (see Muscle index, pp. xx–xx).
- **Arteries**: profunda femoris (p. xx) as well as its medial circumflex femoral and perforating branches and the obturator artery.
- **Veins**: profunda femoris and obturator veins.
- **Nerves**: the anterior and posterior divisions of the obturator nerve (p. xx).

The contents of the posterior compartment of the thigh (Fig. 52.3)
- **Muscles**: these are the *hamstrings* and effect knee flexion and hip extension. They include: *biceps femoris, semitendinosus, semimembranosus* and the *hamstring part of adductor magnus* (see Muscle index, p. xx).
- **Arteries**: the perforating branches of profunda femoris.
- **Veins**: the venae comitantes of the small arteries.
- **Nerves**: the sciatic nerve (L4, L5, S1, S2 and S3, p. xx). The muscles of the posterior compartment are supplied by the tibial component of the sciatic nerve, with the exception of the short head of biceps femoris which is supplied by the common peroneal component.

The adductor (subsartorial or Hunter's) canal

The adductor canal serves to transmit structures from the apex of the femoral triangle through the hiatus in adductor magnus into the popliteal fossa. It commences in the mid-portion of the thigh and is formed by the following walls:
- **The posterior wall**: adductor longus, with adductor magnus in the lower part of the thigh.
- **The lateral wall**: vastus medialis.
- **The roof**: thickened fascia underlying sartorius.

The contents of the adductor canal

These include: the femoral artery, the femoral vein which lies deep to the femoral artery, lymphatics, the saphenous branch of the femoral nerve (which passes behind sartorius to leave the canal and descends the lower limb with the great saphenous vein), the nerve to vastus medialis (in the upper part) and the *subsartorial plexus*. This plexus is formed by branches from the saphenous nerve (terminal branch of the

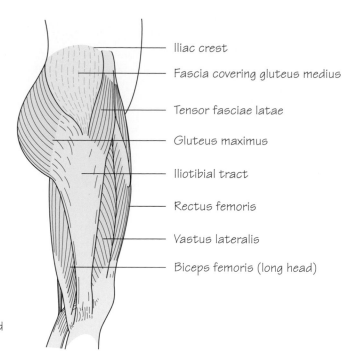

Fig.52.4
The lateral side of the thigh.
Note the two muscles inserted
into the iliotibial tract

Iliac crest

Fascia covering gluteus medius

Tensor fasciae latae

Gluteus maximus

Iliotibial tract

Rectus femoris

Vastus lateralis

Biceps femoris (long head)

femoral nerve, p. xx), the anterior division of the obturator nerve and the intermediate cutaneous nerve of the thigh (branch of the femoral nerve, p. xx). It supplies the skin over the medial aspect of the knee.

Clinical notes

- **Swellings in the groin**: swellings in the groin are a common presenting symptom. Some of the possible causes are:
- Enlarged lymph nodes, either as a result of systemic disease or of infection or malignant tumours in the area of drainage of the femoral lymph nodes.

- Femoral hernia, the sac of which, having traversed the femoral canal, emerges through the saphenous opening. See Chapter 57.
- Inguinal hernia. See Chapter 26.
- A varicose condition of the termination of the great saphenous vein (*saphena varix*).
- Psoas abscess. Tuberculous disease in the lumbar vertebrae may spread into the psoas sheath and thence under the inguinal ligament to present as a swelling in the femoral triangle.
- Enlargement of the bursa that separates the psoas tendon from the hip joint.

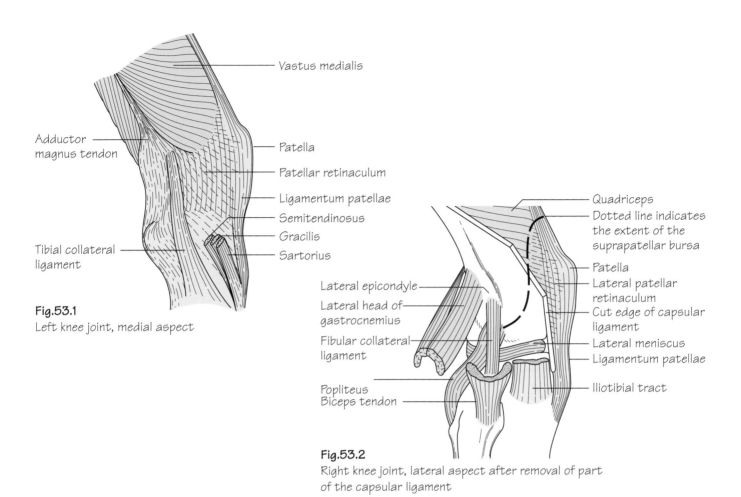

Fig.53.1
Left knee joint, medial aspect

- Vastus medialis
- Adductor magnus tendon
- Patella
- Patellar retinaculum
- Ligamentum patellae
- Semitendinosus
- Gracilis
- Sartorius
- Tibial collateral ligament

Fig.53.2
Right knee joint, lateral aspect after removal of part of the capsular ligament

- Quadriceps
- Dotted line indicates the extent of the suprapatellar bursa
- Patella
- Lateral patellar retinaculum
- Cut edge of capsular ligament
- Lateral meniscus
- Ligamentum patellae
- Iliotibial tract
- Lateral epicondyle
- Lateral head of gastrocnemius
- Fibular collateral ligament
- Popliteus
- Biceps tendon

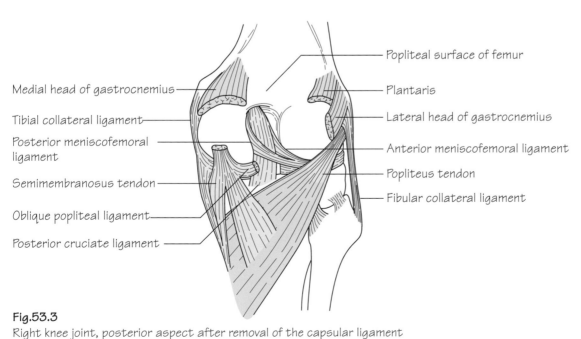

Fig.53.3
Right knee joint, posterior aspect after removal of the capsular ligament

- Popliteal surface of femur
- Plantaris
- Lateral head of gastrocnemius
- Anterior meniscofemoral ligament
- Popliteus tendon
- Fibular collateral ligament
- Medial head of gastrocnemius
- Tibial collateral ligament
- Posterior meniscofemoral ligament
- Semimembranosus tendon
- Oblique popliteal ligament
- Posterior cruciate ligament

Anatomy at a Glance, Third Edition. Omar Faiz, Simon Blackburn and David Moffat.
© 2011 Blackwell Publishing Ltd. Published 2011 by Blackwell Publishing Ltd.

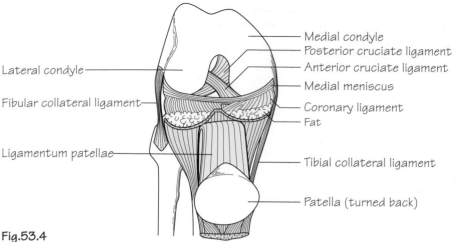

Lateral condyle

Fibular collateral ligament

Ligamentum patellae

Medial condyle
Posterior cruciate ligament
Anterior cruciate ligament
Medial meniscus
Coronary ligament
Fat

Tibial collateral ligament

Patella (turned back)

Fig.53.4
Anterior view of the flexed right knee joint after division of the quadriceps
and retraction of the patella

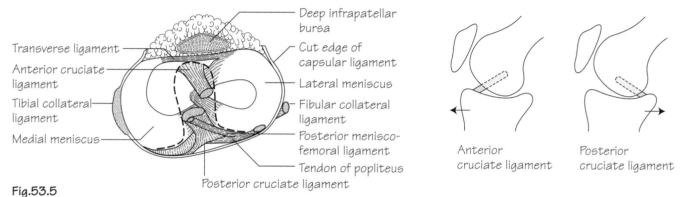

Transverse ligament

Anterior cruciate ligament

Tibial collateral ligament

Medial meniscus

Deep infrapatellar bursa

Cut edge of capsular ligament

Lateral meniscus

Fibular collateral ligament

Posterior menisco-femoral ligament

Tendon of popliteus

Posterior cruciate ligament

Anterior cruciate ligament

Posterior cruciate ligament

Fig.53.5
The upper surface of the tibia and related structures.
The dotted line indicates the synovial membrane in the vicinity of the cruciate ligaments.
The small diagrams show how the cruciate ligaments resist forward and backward displacement of the femur

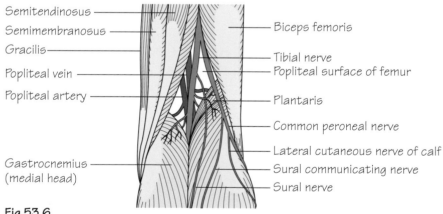

Semitendinosus

Semimembranosus

Gracilis

Popliteal vein

Popliteal artery

Gastrocnemius (medial head)

Biceps femoris

Tibial nerve
Popliteal surface of femur

Plantaris

Common peroneal nerve

Lateral cutaneous nerve of calf
Sural communicating nerve
Sural nerve

Fig.53.6
The right popliteal fossa

The knee joint (Figs. 53.1–53.5)

- **Type**: it is a *synovial modified hinge joint* which permits a small degree of rotation. In the knee joint, the femoral and tibial condyles articulate as does the patella and patellar surface of the femur. Note that the fibula does *not* contribute to the knee joint.
- **Capsule**: the articular surfaces are covered by articular cartilage. The capsule is attached to the margins of the articular surfaces, except anteriorly where it dips downwards. In the anterior part of the capsule, there is a large opening through which the synovial membrane is continuous with the *suprapatellar bursa* (Fig. 53.2). This bursa extends superiorly three fingers' breadth above the patella between the femur and quadriceps. Posteriorly, the capsule communicates with another *bursa* under the *medial head of gastrocnemius* and often, through it, with the *bursa of semimembranosus*. Posterolaterally, another opening in the capsule permits the passage of the tendon of *popliteus*.
- **Extracapsular ligaments**: the capsule of the knee joint is reinforced by ligaments.
  - The *medial (tibial) collateral ligament* (Figs. 53.1 and 53.3): consists of superficial and deep parts. The superficial component is attached above to the femoral epicondyle and below to the subcutaneous surface of the tibia. The deep component is firmly attached to the medial meniscus.
  - The *lateral (fibular) collateral ligament* (Fig. 53.2): is attached to the femoral epicondyle above and, along with biceps femoris, to the head of the fibula below. Unlike the medial collateral ligament it lies away from the capsule and meniscus.

  The collateral ligaments are taut in full extension, and it is in this position that they are liable to injury when subjected to extreme valgus/varus strain.

  Behind the knee, the *oblique popliteal ligament*, a reflected extension from the semimembranosus tendon, strengthens the capsule (Fig. 53.3). Anteriorly, the capsule is reinforced by the *ligamentum patellae* and the *patellar retinacula*. The latter are reflected fibrous expansions arising from vastus lateralis and medialis muscles which blend with the capsule anteriorly (Fig. 53.1).
- **Intracapsular ligaments**: the *cruciate ligaments* are enclosed within the knee joint (Figs. 53.4 and 53.5).
  - The *anterior cruciate ligament*: passes from the front of the intercondylar area of the tibia to the medial side of the lateral femoral condyle. This ligament prevents hyperextension and resists forward movement of the tibia on the femur.
  - The *posterior cruciate ligament*: passes from the back of the intercondylar area of the tibia to the lateral side of the medial condyle. It becomes taut in hyperflexion and resists posterior displacement of the tibia on the femur.
- **The menisci (semilunar cartilages)**: these are crescentic fibrocartilaginous 'shock absorbers' within the joint. They lie within deepened grooves on the articular surfaces of the tibial condyles (Fig. 53.5). The *medial meniscus* is C-shaped and larger than the *lateral meniscus*. The menisci are attached to the tibial intercondylar area by their horns and around their periphery by small coronary ligaments. The lateral meniscus is loosely attached to the tibia and connected to the femur by two *meniscofemoral ligaments* (see Fig. 53.3).
- **Blood supply**: is from the rich anastomosis formed by the genicular branches of the popliteal artery.

- **Nerve supply**: is from branches of the femoral, tibial, common peroneal and obturator nerves.

Knee movements

Flexion and extension are the principal movements at the knee. Some rotation is possible when the knee is flexed but this is lost in extension. During the terminal stages of extension the large medial tibial condyle screws forwards onto the femoral condyle to lock the joint. Conversely, the first stage of flexion is unlocking the joint by internal rotation of the medial tibial condyle—an action performed by popliteus.

The principal muscles acting on the knee are:
- **Extension**: quadriceps femoris.
- **Flexion**: predominantly the hamstrings but also gracilis, gastrocnemius and sartorius.
- **Rotation**: popliteus effects internal (medial) rotatory movement of the tibia.

The popliteal fossa (Fig. 53.6)

The femoral artery and vein pass through the hiatus in adductor magnus to enter the popliteal fossa and, in so doing, become the popliteal vessels.

The popliteal fossa is rhomboidal in shape. Its superior boundaries are the biceps tendon (superolateral) and semimembranosus reinforced by semitendinosus (superomedial). The medial and lateral heads of gastrocnemius form the inferomedial and inferolateral boundaries, respectively.
- **The roof consists of** the deep fascia which is penetrated at an inconstant position by the small saphenous vein as it drains into the popliteal vein.
- **The floor consists of (from above downwards)** the posterior lower femur, the posterior surface of the knee joint and popliteus.
- **The contents of the fossa include (from deep to superficial)** the *popliteal artery*, *vein* and the *tibial nerve*. The *common peroneal nerve* runs along the medial border of biceps tendon and then out of the fossa. Other contents include fat and *popliteal lymph nodes*.

The popliteal pulse is notoriously difficult to feel because the artery lies deep to other structures. Whenever a popliteal pulse is easily palpable, the possibility of aneurysmal change should be considered.

Clinical notes

- **Meniscus injury**: the menisci are especially prone to flexion/rotation injuries of the knee, the medial meniscus being the more vulnerable because it is firmly attached to the medial ligament and is, therefore, less mobile. The classic medial meniscus injury occurs when a footballer twists the knee during running exerting a combination of external rotation and abduction with the joint flexed. The most common type of injury is the *bucket-handle* tear in which the meniscus splits along its length.
- **Cruciate ligament tears**: rupture of the cruciate ligaments leads to a very unstable knee joint in which the tibia can be displaced backwards and forwards on the femur.

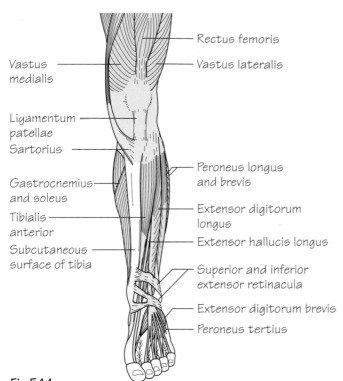

Rectus femoris

Vastus lateralis

Vastus medialis

Ligamentum patellae

Sartorius

Gastrocnemius and soleus

Tibialis anterior

Subcutaneous surface of tibia

Peroneus longus and brevis

Extensor digitorum longus

Extensor hallucis longus

Superior and inferior extensor retinacula

Extensor digitorum brevis

Peroneus tertius

Fig.54.1
The extensor (dorsiflexor) group of muscles

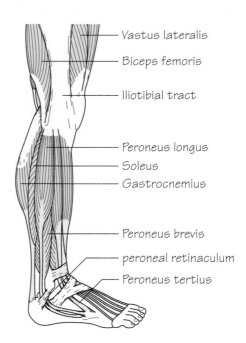

Vastus lateralis

Biceps femoris

Iliotibial tract

Peroneus longus

Soleus

Gastrocnemius

Peroneus brevis

peroneal retinaculum

Peroneus tertius

Fig.54.2
The lateral side of the leg and foot

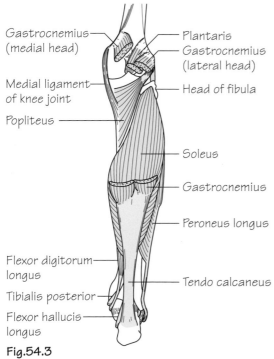

Gastrocnemius (medial head)

Plantaris

Gastrocnemius (lateral head)

Medial ligament of knee joint

Head of fibula

Popliteus

Soleus

Gastrocnemius

Peroneus longus

Flexor digitorum longus

Tendo calcaneus

Tibialis posterior

Flexor hallucis longus

Fig.54.3
The superficial muscles of the calf

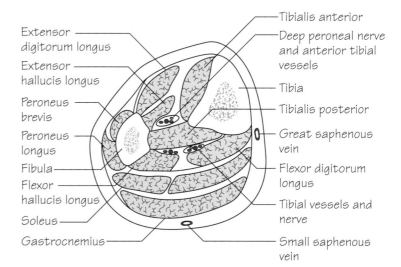

Extensor digitorum longus

Extensor hallucis longus

Peroneus brevis

Peroneus longus

Fibula

Flexor hallucis longus

Soleus

Gastrocnemius

Tibialis anterior

Deep peroneal nerve and anterior tibial vessels

Tibia

Tibialis posterior

Great saphenous vein

Flexor digitorum longus

Tibial vessels and nerve

Small saphenous vein

Fig.54.4
Cross-section through the leg, to be studied in conjunction with Figs.46.2 and 46.3

Within the leg, there are three predominant muscle groups: extensor, peroneal and flexor. Each of these groups has an individual blood and nerve supply.

Students are often confused about the description of movements of the foot. *Extension* of the foot (dorsiflexion) refers to lifting the toes and the ball of the foot upwards. Conversely, foot *flexion* (plantarflexion) is the opposing action.

The deep fascia of the leg

The deep fascia of the leg is continuous above with the deep fascia of the thigh. It envelops the leg and fuses with the periosteum of the tibia at the anterior and medial borders. Other fascial septa, and the interosseous membrane, divide the leg into four compartments: *extensor*, *peroneal*, *superficial flexor* and *deep flexor*.

The superior and inferior tibiofibular joints

These are, respectively, synovial and fibrous joints between the tibia and fibula at their proximal and distal ends.

The interosseous membrane (Fig. 54.4)

The interosseous borders of the tibia and fibula are connected by a strong sheet of connective tissue – the *interosseous membrane*. The fibres of the membrane run obliquely downwards from tibia to fibula. Its function is to bind together the bones of the leg as well as providing a surface for muscle attachment.

The extensor aspects of the leg and dorsum of the foot (Figs. 54.1 and 54.4)

The extensor group consists of four muscles in the leg (see below) and extensor digitorum brevis in the foot. These muscles dorsiflex the foot and toes. The contents of the extensor compartment of the leg are as follows:
- **Muscles**: tibialis anterior, extensor hallucis longus, extensor digitorum longus and peroneus tertius (unimportant in function) (see Muscle index, p. xx).
- **Artery**: the anterior tibial artery (p. xx) and its venae comitantes form the vascular supply of the extensor compartment. The artery continues as the dorsalis pedis artery in the foot.
- **Nerves**: the deep peroneal nerve (p. xx) supplies all of the muscles of the extensor compartment. Injury to this nerve results in the inability to dorsiflex the foot – foot drop (see Chapter 50).

The extensor retinacula (Fig. 54.1)

These are thickenings of the deep fascia of the leg. They serve to stabilise the underlying extensor tendons.
- The *superior extensor retinaculum* is a transverse band attached to the anterior borders of the tibia and fibula.
- The *inferior extensor retinaculum* is Y-shaped. Medially, the two limbs attach to the medial malleolus and the plantar aponeurosis and, laterally, the single limb is attached to the calcaneus.

The peroneal compartment of the leg
(Figs. 54.2 and 54.4)

This compartment consists of two muscles – peroneus longus and brevis. These muscles are the predominant foot everters. The contents of the peroneal compartment include:
- **Muscles**: *peroneus longus and brevis* (see Muscle index, p. xx).
- **Artery**: the *peroneal artery* (p. xx).
- **Nerve**: the superficial peroneal nerve (p. xx).

Peroneal retinacula (Fig. 54.2)

The *superior peroneal retinaculum* is a thickening of deep fascia attached from the lateral malleolus to the calcaneus. The *inferior peroneal retinaculum* is a similar band of fascia which is continuous with the inferior extensor retinaculum. The tendons of peroneus longus and brevis pass in their synovial sheaths beneath.

The flexor aspect of the leg (Fig. 54.2)

The flexor muscles of the calf are considered in superficial and deep groups. All flexor muscles of the calf receive their nerve and arterial supplies from the tibial nerve and the posterior tibial artery, respectively.

The contents of the flexor compartment of the calf include:
- **Superficial flexor muscle group**: *gastrocnemius*, *soleus* and *plantaris* (the last is rudimentary in humans). Note that all of these muscles are inserted into the middle third of the posterior surface of the calcaneus via the *tendo calcaneus* (*Achilles tendon*). A small bursa (the *retrocalcaneal bursa*) occupies the space between the upper third of the posterior surface of the calcaneus and the Achilles tendon. Within soleus and, to a lesser extent, gastrocnemius, there is an extensive venous plexus. These muscles act as a muscle pump, squeezing venous blood upwards during their contraction. It is in these veins that deep venous thromboses readily occur after surgery in the immobile patient (p. xx).
- **Deep flexor muscle group**: *tibialis posterior*, *flexor digitorum longus*, *flexor hallucis longus* (see Muscle index, p. xx).
- **Artery**: *posterior tibial artery* (p. xx).
- **Nerve**: *tibial nerve* (p. xx).

Clinical notes

- **Compartment syndrome**: following the fractures of the leg, oedema, within one or more compartments, can lead to obstruction to blood flow with consequent failure of blood supply to the occupying tissues – *compartment syndrome*. When this occurs, immediate decompression (*fasciotomy*) of all four compartments is necessary.
- **Fractures of the tibia**: reference to the Muscle index will show that there are no muscles attached to the lower third of the tibia. As the muscle vasculature is the main source of blood supply to the periosteum and, thence, to the bone, fractures in this region are especially prone to delayed healing.
- **Bursitis of the tendo calcaneus**: the bursa between the tendo calcaneus and the back of the calcaneus may become inflamed, producing an extremely painful condition, sometimes caused by the pressure of new shoes.

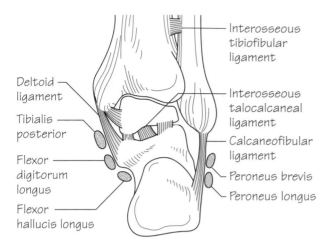

Interosseous tibiofibular ligament

Deltoid ligament

Interosseous talocalcaneal ligament

Tibialis posterior

Calcaneofibular ligament

Flexor digitorum longus

Peroneus brevis

Flexor hallucis longus

Peroneus longus

Fig.55.1
The ankle joint from behind, to show how the talus is held in position by ligaments between the tibia and fibula above and the calcaneus below

Joints
1. Subtalar joint
2. Midtarsal (talonavicular)
3. Tarsometatarsal
4. Metatarsophalangeal
5. Midtarsal (calcaneocuboid)

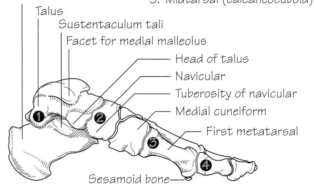

Calcaneus

Talus

Sustentaculum tali

Facet for medial malleolus

Head of talus

Navicular

Tuberosity of navicular

Medial cuneiform

First metatarsal

Sesamoid bone

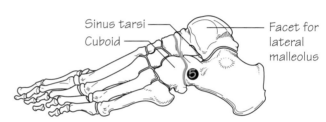

Sinus tarsi

Cuboid

Facet for lateral malleolus

Fig.55.2
The bones of the foot, medial and lateral aspects. The major joints are shown

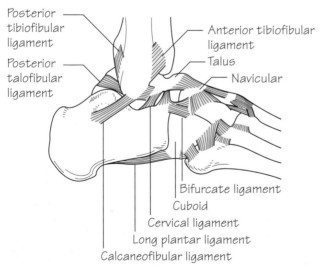

Posterior tibiofibular ligament

Anterior tibiofibular ligament

Posterior talofibular ligament

Talus

Navicular

Bifurcate ligament

Cuboid

Cervical ligament

Long plantar ligament

Calcaneofibular ligament

Fig.55.3
The ankle joint, lateral aspect after removal of the capsular ligament

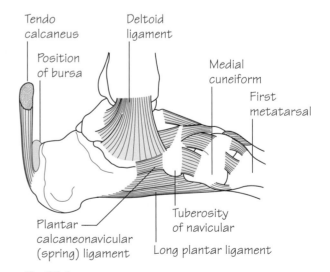

Tendo calcaneus

Deltoid ligament

Position of bursa

Medial cuneiform

First metatarsal

Plantar calcaneonavicular (spring) ligament

Tuberosity of navicular

Long plantar ligament

Fig.55.4
The ankle joint, medial aspect

The ankle joint (Fig. 55.1)

- **Type**: the ankle is a *synovial hinge joint* involving the tibia, fibula and talus. The articular surfaces are covered with cartilage and synovial membrane lines the rest of the joint.
- **Capsule**: the capsule encloses the articular surfaces. The capsule is reinforced on either side by strong *collateral ligaments*, but is lax anteriorly to permit uninhibited hinged movement.
- **Ligaments**: the *medial collateral (deltoid) ligament* consists of a deep component, which is a vertical band passing from the medial malleolus to the talus and a superficial component, which is fan-shaped. The superficial component extends from the medial malleolus to (from front to back): the tuberosity of the navicular, the spring ligament (see below), the sustentaculum tali and the posterior tubercle of the talus (Figs. 55.1 and 55.4).

The *lateral collateral ligament* consists of three bands: the anterior and posterior talofibular ligaments and the calcaneofibular ligament (Fig. 55.3).

The movements at the ankle

It is important to note that inversion and eversion movements of the foot do not occur at the ankle joint except in full plantarflexion. These occur at the *subtalar* and *midtarsal* joints (see below). Only dorsiflexion (extension) and plantarflexion (flexion) occur at the ankle. The principal muscles are:

- **Dorsiflexion**: tibialis anterior and, to a lesser extent, extensor hallucis longus and extensor digitorum longus.
- **Plantarflexion**: gastrocnemius and soleus and, to a lesser extent, tibialis posterior, flexor hallucis longus and flexor digitorum longus.

The foot bones (Fig. 55.2)

With the exception of the metatarsals and phalanges, the foot bones are termed collectively the *tarsal bones*.

- **Talus**: has a body with facets on the superior, medial and lateral surfaces for articulation with the tibia, medial malleolus and lateral malleolus, respectively. There is a groove on the posterior surface of the body for the tendon of flexor hallucis longus. To the groove's lateral side is the posterior (lateral) tubercle, sometimes known as the *os trigonum*, as it ossifies from a separate centre to the talus. A head projects distally, which articulates with the navicular. The head is connected to the body by a neck.
- **Calcaneus**: has two facets on the superior surface which participate in the *subtalar* (talocalcaneal and talocalcaneonavicular) joint. The posterior surface has three areas: a roughened middle part where the tendo calcaneus inserts, a smooth upper part which is separated from the tendo calcaneus by a bursa (retrocalcaneal bursa) (Fig. 55.4) and a lower part which is covered by a fibro-fatty pad that forms the heel. Medial and lateral tubercles are present on the inferior surface to which the plantar aponeurosis is attached. The sustentaculum tali is a distinctive projection on the medial surface which forms a shelf for the support of the talus. The *peroneal tubercle*, a small projection on the lateral surface of the calcaneus, separates the tendons of peroneus longus and brevis. The anterior surface has a facet for articulation with the cuboid.
- **Cuboid**: has a grooved undersurface for the tendon of peroneus longus.

- **Navicular**: has facets for the articulations with the head of the talus posteriorly and the three cuneiforms anteriorly. It has a tuberosity on its medial aspect which provides attachment for tibialis posterior.
- **Cuneiforms**: there are three cuneiforms which articulate anteriorly with the metatarsals and posteriorly with the navicular. Their wedge shape helps to maintain the *transverse arch* of the foot.
- **Metatarsals and phalanges**: these are similar to the metacarpals and phalanges of the hand. Note the articulations of the heads of the metatarsals. The 1st metatarsal is large and is important for balance. The head is grooved on its inferior surface for the two sesamoid bones within the tendon of flexor hallucis brevis.

The foot joints

- **Subtalar joint (Fig. 55.2)**: this compound joint comprises the *talocalcaneal* and the *talocalcaneonavicular* joints. Inversion and eversion movements occur at the subtalar joint.
  - The *talocalcaneal joint*: is a synovial plane joint formed by the articulation of the upper surface of the calcaneus with the lower surface of the talus.
  - The *talocalcaneonavicular joint*: is a synovial ball and socket joint between the head of the talus and the sustentaculum tali, the spring ligament and the navicular.
- **Midtarsal joint (Fig. 55.2)**: is also a compound joint which contributes towards foot inversion/eversion movements. This joint is composed of the *calcaneocuboid* joint and the *talonavicular* component of the *talocalcaneonavicular* joint.
  - The *calcaneocuboid joint*: is a synovial plane joint formed between the anterior surface of the calcaneus and the posterior surface of the cuboid.
- **Other foot joints (Fig. 55.2)**: these include other tarsal joints, tarsometatarsal (synovial plane), intermetatarsal (synovial plane), metatarsophalangeal (synovial condyloid) and interphalangeal (synovial hinge) joints.

Clinical notes

- **Injuries to the ankle**: sudden inversion or eversion of the weight-bearing foot may lead to tearing of the medial or lateral ligaments. If complete tear occurs, the talus will not be held firmly between the malleoli, and the ankle joint will be unstable. More severe injuries, especially if combined with rotational strain, will lead to malleolar fracture, usually of the lateral malleolus, with tearing of the medial ligament, which may avulse the medial malleolus itself. In the most severe form of these injuries (Pott's fractures), the leg is displaced forward on the foot with fracture of the posterior lip of the tibial facet (sometimes called the *third malleolus*).
- **March fractures**: the second metatarsal is more slender than the others and projects further forward (Fig. 55.2). For this reason, if the intrinsic muscles of the foot are weakened, for example by fatigue in soldiers during a long march, the first metatarsal no longer carries the main weight of the body and the second metatarsal may fracture spontaneously.

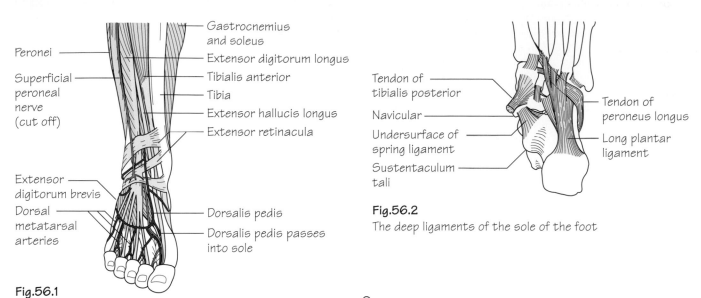

Peronei

Superficial peroneal nerve (cut off)

Extensor digitorum brevis

Dorsal metatarsal arteries

Gastrocnemius and soleus

Extensor digitorum longus

Tibialis anterior

Tibia

Extensor hallucis longus

Extensor retinacula

Dorsalis pedis

Dorsalis pedis passes into sole

Fig.56.1
The structures on the front of the ankle and the dorsum of the foot

Tendon of tibialis posterior

Navicular

Undersurface of spring ligament

Sustentaculum tali

Tendon of peroneus longus

Long plantar ligament

Fig.56.2
The deep ligaments of the sole of the foot

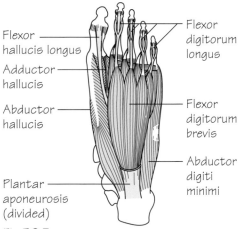

Flexor hallucis longus

Adductor hallucis

Abductor hallucis

Plantar aponeurosis (divided)

Flexor digitorum longus

Flexor digitorum brevis

Abductor digiti minimi

Fig.56.3
The first layer of muscles in the sole of the foot

Flexor hallucis brevis

Abductor hallucis

Flexor hallucis longus

Flexor digitorum longus

Plantar aponeurosis

2nd and 3rd lumbricals

Abductor digiti minimi

Flexor digiti minimi brevis

Flexor digitorum accessorius

Flexor digitorum brevis

Fig.56.4
The second layer of muscles in the sole of the foot

Sesamoid bones

Abductor hallucis

Flexor hallucis brevis

Tibialis posterior tendon

Transverse and oblique heads of adductor hallucis

Flexor digiti minimi brevis

Fibrous covering of peroneus longus tendon

Fig.56.5
The third layer of muscles in the sole of the foot

 Anatomy at a Glance, Third Edition. Omar Faiz, Simon Blackburn and David Moffat.
© 2011 Blackwell Publishing Ltd. Published 2011 by Blackwell Publishing Ltd.

Ligaments of the foot

- **Spring (plantar calcaneonavicular) ligament (Fig. 56.2)**: it runs from the sustentaculum tali to the tuberosity of the navicular forming a support for the head of the talus.
- **Bifurcate ligament**: is Y-shaped and runs from the anterior part of the calcaneus to the cuboid and navicular bones. It reinforces the capsule of the talocalcaneonavicular joint.
- **Long plantar ligament (Figs. 55.4 and 56.2)**: It runs from the undersurface of the calcaneus to the cuboid and bases of the lateral metatarsals. The ligament runs over the tendon of peroneus longus.
- **Short plantar ligament**: runs from the undersurface of the calcaneus to the cuboid.
- **Medial and lateral (talocalcaneal) ligaments**: strengthen the capsule of the talocalcaneal joint.
- **Interosseous talocalcaneal ligament**: runs in the sinus tarsi, a tunnel formed by deep sulci on the talus and calcaneus.
- **Deep transverse metatarsal ligaments**: join the plantar ligaments of the metatarsophalangeal joints of the five toes.

The arches of the foot

The integrity of the foot is maintained by two *longitudinal* (medial and lateral) arches and a single *transverse* arch. The arches are held together by a combination of bony, ligamentous and muscular factors, so that standing weight is taken on the posterior part of the calcaneum and the metatarsal heads as a result of the integrity of the arches.

- **Medial longitudinal arch (see Fig. 55.2)**: comprises calcaneus, talus (the apex of the arch), navicular, the three cuneiforms and three medial metatarsals. The arch is bound together by the spring ligament and muscles and supported from above by tibialis anterior and posterior.
- **Lateral longitudinal arch (see Fig. 55.2)**: comprises calcaneus, cuboid and the two lateral metatarsals. The arch is bound together by the long and short plantar ligaments and supported from above by peroneus longus and brevis.
- **Transverse arch**: comprises the cuneiforms and bases of the metatarsals. The arch is bound together by the deep transverse ligament, plantar ligaments and the interossei. It is supported from above by peroneus longus and brevis.

The dorsum of the foot (Fig. 56.1)

The skin of the dorsum of the foot is supplied by cutaneous branches of the superficial peroneal, deep peroneal, saphenous and sural nerves. The dorsal venous arch lies within the subcutaneous tissue overlying the metatarsal heads. It receives blood from most of the superficial tissues of the foot via digital and communicating branches. The great saphenous vein commences from the medial end of the arch and the small saphenous vein from the lateral end.

Structures on the dorsum of the foot (Fig. 56.1)

- **Muscles**: extensor digitorum brevis arises from the calcaneus. Other muscles insert on the dorsum of the foot but arise from the leg. These include tibialis anterior, extensor hallucis longus, extensor digitorum longus, peroneus tertius and peroneus brevis. Each tendon of extensor digitorum longus is joined on its lateral side by a tendon from extensor digitorum brevis. The tendons of extensor digitorum longus and peroneus tertius share a common synovial sheath, whilst the other tendons have individual sheaths.
- **Arterial supply**: is from the dorsalis pedis artery – the continuation of the anterior tibial artery. The dorsalis pedis ends by passing to the sole where it completes the plantar arch (p. xx).
- **Nerve supply**: is from the deep peroneal nerve via its medial and lateral terminal branches. The latter supplies extensor digitorum brevis, whereas the former receives cutaneous branches from the skin.

The sole of the foot

The sole is described as consisting of an aponeurosis and four muscle layers. The skin of the sole is supplied by the medial and lateral plantar branches of the tibial nerve. The medial calcaneal branch of the tibial nerve innervates a small area on the medial aspect of the heel.

The plantar aponeurosis

This aponeurosis lies deep to the superficial fascia of the sole and covers the 1st layer of muscles. It is attached to the calcaneus behind and sends a deep slip to each toe as well as blending superficially with the skin creases at the base of the toes. The slips that are sent to each toe split into two parts which pass on either side of the flexor tendons and fuse with the *deep transverse metatarsal ligaments*.

The muscular layers of the sole

- **1st layer consists of**: abductor hallucis, flexor digitorum brevis and abductor digiti minimi (Fig. 56.3).
- **2nd layer consists of**: flexor digitorum accessorius, the lumbricals and the tendons of flexor digitorum longus and flexor hallucis longus (Fig. 56.4).
- **3rd layer consists of**: flexor hallucis brevis, adductor hallucis and flexor digiti minimi brevis (Fig. 56.5).
- **4th layer consists of**: the dorsal and plantar interossei and the tendons of peroneus longus and tibialis posterior.

Neurovascular structures of the sole

- **Arterial supply**: is from the posterior tibial artery which divides into *medial* and *lateral plantar* branches. The latter branch contributes the major part of the *deep plantar arch* (p. xx).
- **Nerve supply**: is from the tibial nerve which also divides into *medial* and *lateral plantar* branches (p. xx).

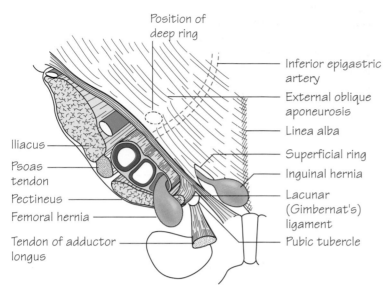

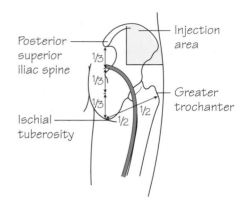

Fig.57.1
The anatomy of femoral and inguinal herniae.
Note the relation of the deep inguinal ring
to the inferior epigastric artery and the
relation of the two types of hernia to the
pubic tubercle

Fig.57.2
The surface markings of the sciatic nerve

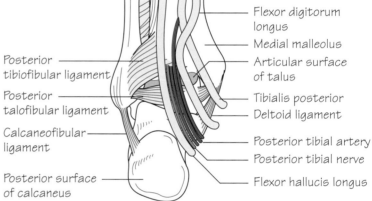

Fig.57.3
The structures on the medial side of the ankle

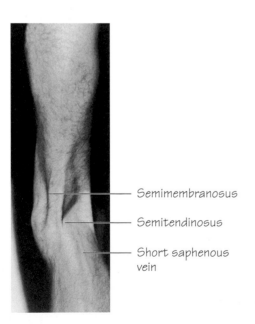

Fig.57.4
Visible structures on the medial side
of the lower limb

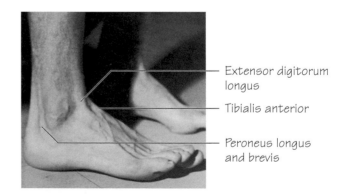

Fig.57.5
The lateral aspect of the foot to show
the tendons that can be recognized

Surface landmarks around the hip and gluteal region

- **The anterior superior iliac spine** is a prominent landmark at the anterior end of the iliac crest.
- **The greater trochanter** of the femur lies approximately a hand's breadth below the iliac crest. It is made more prominent by adducting the hip.
- **The ischial tuberosity** is covered by gluteus maximus when the hip is extended. It can be palpated in the lower part of the buttock with the hip flexed.
- **The femoral pulse** (Fig. 57.1) is most easily felt halfway between the anterior superior iliac spine and the symphysis pubis (mid-inguinal point). The femoral head lies deep to the femoral artery at the mid-inguinal point. The femoral vein lies medial, and the femoral nerve lateral, to the artery at this point.
- **The femoral canal** (Fig. 57.1) lies medial to the femoral vein within the femoral sheath.
- **The great saphenous vein** pierces the cribriform fascia in the saphenous opening of the deep fascia to drain into the femoral vein 4 cm below and lateral to the pubic tubercle (Fig. 48.2).
- In thin subjects, the *horizontal chain* of **superficial inguinal lymph nodes** is palpable. It lies below and parallel to the inguinal ligament.
- **The sciatic nerve** has a curved course throughout the gluteal region. Consider two lines—one connects the posterior superior iliac spine and the ischial tuberosity and the other connects the greater trochanter and the ischial tuberosity (Fig. 57.2). The sciatic nerve crosses the first line at the junction of the superior 1/3 and inferior 2/3, then crosses the second line at its midpoint. The nerve then descends the thigh in the midline posteriorly. The division of the sciatic nerve into tibial and common peroneal components occurs usually at a point a hand's breadth above the popliteal crease, but is highly variable.
- **The common peroneal nerve** winds superficially around the neck of the fibula. In thin subjects, it can be palpated at this point.

Surface landmarks around the knee
(Fig. 57.4)

- **The patella and ligamentum patellae** are easily palpable with the limb extended and relaxed. The ligamentum patellae can be traced to its attachment at the *tibial tuberosity*.
- **The adductor tubercle** can be felt on the medial aspect of the femur above the medial condyle.
- **The femoral and tibial condyles** are prominent landmarks. With the knee in flexion, the joint line and outer edges of the *menisci* within are palpable. The *medial* and *lateral collateral ligaments* are palpable on either side of the knee and can be followed to their bony attachments.
- **The subcutaneous border of the tibia** is palpable throughout its length.
- **The fibular head** is palpable laterally. The shaft of the fibula is mostly covered but is subcutaneous for the terminal 10 cm.
- **The popliteal pulse** is difficult to feel as it lies deep to the tibial nerve and popliteal vein. It is best felt by palpating in the popliteal fossa with the patient prone and the knee flexed.

Surface landmarks around the ankle

- The medial and lateral malleoli are prominent at the ankle. The lateral is more elongated and descends a little further than the medial.

- When the foot is dorsiflexed, the tendons of *tibialis anterior*, *extensor hallucis longus* and *extensor digitorum* are visible on the anterior aspect of the ankle and the dorsum of the foot.
- The tendons of *peroneus longus* and *brevis* pass behind the lateral malleolus.
- Passing behind the medial malleolus lie: the tendons of *tibialis posterior* and *flexor digitorum longus*, the *posterior tibial artery* and its *venae comitantes*, the *tibial nerve* and *flexor hallucis longus* (Fig. 57.3).

Surface landmarks around the foot
(Fig. 57.5)

- **The head of the talus** is palpable immediately anterior to the distal tibia.
- **The base of the 5th metatarsal** is palpable on the lateral border of the foot. The tendon of peroneus brevis inserts onto the tuberosity on the base.
- **The heel** is formed by the calcaneus. The *tendo calcaneus* (*Achilles tendon*) is palpable above the heel. Sudden stretch of this can lead to rupture. When this occurs, a gap in the tendon is often palpable.
- **The tuberosity of the navicular** can be palpated 2.5 cm anterior to the medial malleolus. It receives most of the tendon of tibialis posterior.
- **The peroneal tubercle** of the calcaneum can be felt 2.5 cm below the tip of the lateral malleolus.
- **The sustentaculum tali** can be felt 2.5 cm below the medial malleolus. The tendon of tibialis posterior lies above the sustentaculum tali and the tendon of flexor hallucis longus winds beneath it.
- **The dorsalis pedis pulse** is located on the dorsum of the foot between the tendons of extensor hallucis longus and extensor digitorum.
- **The posterior tibial pulse** is best felt halfway between the medial malleolus and the heel.
- **The dorsal venous arch** is visible on the dorsum of the foot. The small saphenous vein drains the lateral end of the arch and passes posterior to the lateral malleolus to ascend the calf and drain into the popliteal vein. The great saphenous vein passes anterior to the medial malleolus to ascend the length of the lower limb and drain into the femoral vein.

Clinical notes

- The great saphenous vein is often visible just above and in front of the medial malleolus, but it can always be found here, even if it cannot be seen, by making a small incision under local anaesthesia. This is useful when intravenous access is required urgently and other veins are difficult to find.
- The sac of a femoral hernia passes through the canal to expand below the deep fascia. The hernial sac always lies below and lateral to the pubic tubercle (cf. the neck of an inguinal hernia which is always situated above and medial to the tubercle – Fig. 57.1). The risk of strangulation is high in femoral herniae as the femoral canal is narrow and blood flow to viscera within the hernial sac can easily be impaired.

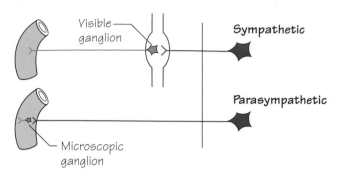

Fig.58.1

The different lengths of the pre- and postganglionic fibres of the autonomic nervous system.
Preganglionic fibres: red
Postganglionic fibres: green

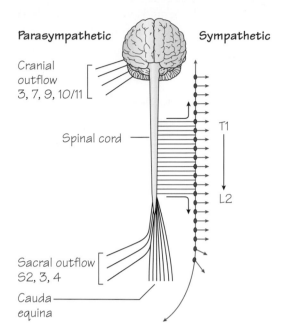

Fig.58.2

The sympathetic (right) and parasympathetic (left) outflows

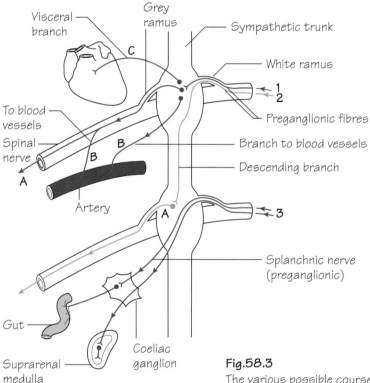

Fig.58.3

The various possible courses of the preganglionic fibres (1, 2 & 3) and postganglionic fibres (A, B and C) of the sympathetic nervous system

The autonomic nervous system comprises two parts – the *sympathetic* and the *parasympathetic*. The former initiates the 'fight or flight' reaction, while the latter controls the body under more relaxed conditions. Both systems have synapses in peripheral ganglia, but those of the sympathetic system are, for the most part, close to the spinal cord in the ganglia of the *sympathetic trunk*, whereas those of the parasympathetic system are mostly in the walls of the viscera themselves and are microscopic (except for the four macroscopic ganglia in the head and neck described below). Thus, the sympathetic preganglionic fibres are relatively short compared with the parasympathetic fibres (Fig. 58.1).

• **Sympathetic outflow (Fig. 58.2)**: the anterior rami of spinal nerves T1 to L2 or L3. The fibres leave these spinal nerves as the *white rami communicantes* and synapse in the ganglia of the sympathetic trunk.

- **Parasympathetic outflow**: comprises:
  - *Cranial outflow*: fibres travel as 'passengers' in the cranial nerves 3, 7, 9 and 10/11 and synapse in one of the four macroscopic peripheral ganglia of the head and neck.
  - *Sacral outflow*: fibres travel in sacral nerves S2, S3 and S4.

The sympathetic system

- **The sympathetic trunk**: runs from the base of the skull to the tip of the coccyx where the two trunks join to form the *ganglion impar*. The trunk continues upwards into the carotid canal as the *internal carotid nerve*.
- **Superior cervical ganglion**: represents the fused ganglia of C1, C2, C3 and C4.
- **Middle cervical ganglion**: represents the fused ganglia of C5 and C6.
- **Inferior cervical ganglion**: represents the fused ganglia of C7 and C8. It may be fused with the ganglion of T1 to form the *stellate ganglion*.

  For courses of the pre- and postganglionic fibres, see Fig. 58.3.
- **Preganglionic fibres**: when the white (myelinated) rami reach the sympathetic trunk, they may follow one of the three different routes:
  1 They may synapse with a nerve cell in the corresponding ganglion.
  2 They may pass straight through the corresponding ganglion and travel up or down the sympathetic trunk, to synapse in another ganglion.
  3 They may pass straight through their own ganglion, maintaining their preganglionic status until they synapse in one of the outlying ganglia, such as the coeliac ganglion. *One exceptional group of fibres even passes through the coeliac ganglion and does not synapse until it reaches the suprarenal medulla.*
- **Postganglionic fibres**: after synapsing, the postganglionic fibres may follow one of the three different routes:
  1 They may pass back to the spinal nerve as a grey (unmyelinated) ramus and are then distributed with the branches of that nerve.
  2 They may pass to adjacent arteries to form a plexus around them and are then distributed with the branches of the arteries. Other fibres leave branches of the spinal nerves later to pass to the arteries more distally.

3 They may pass directly to the viscera in distinct and sometimes named branches, such as the *cervical cardiac branches* of the cervical ganglia.

If the sympathetic trunk is divided above T1 or below L2, the head and neck or the lower limb will lose all sympathetic supply.

Details of the sympathetic system in the various regions are given in the appropriate chapters, but Table 58.1 summarises the autonomic supply to the most important regions and viscera.

The parasympathetic system

- **The cranial outflow**:

III The *oculomotor nerve* carries parasympathetic fibres to the *constrictor pupillae* and the *ciliary muscle*, synapsing in the *ciliary ganglion*.

VII The *facial nerve* carries fibres for the *submandibular* and *sublingual* glands (which synapse in the *submandibular ganglion*) and for the *lacrimal* gland (which synapse in the *sphenopalatine ganglion*).

IX The *glossopharyngeal nerve* carries fibres for the *parotid* gland, which synapse in the *otic* ganglion.

X/XI The *vagus and cranial root of the accessory* carry fibres for the thoracic and abdominal viscera down as far as the proximal two-thirds of the transverse colon, where supply is taken over by the sacral outflow. Synapses occur in minute ganglia in the cardiac and pulmonary plexuses and in the walls of the viscera.

- **The sacral outflow**:

From the sacral nerves S2, S3 and S4, fibres join the inferior *hypogastric plexuses* by means of the *pelvic splanchnic nerves*. They go on to supply the pelvic viscera, synapsing in minute ganglia in the walls of the viscera themselves. Some fibres climb out of the pelvis around the inferior mesenteric artery and supply the sigmoid and descending colon and the distal one-third of the transverse colon.

Clinical notes

- **Horner's syndrome:** loss of the sympathetic supply to the head and neck will produce *Horner's syndrome*. There will be loss of sweating (*anhidrosis*), drooping of the upper eyelid (*ptosis*) and constriction of the pupil (*myosis*) on that side.

Table 58.1 The autonomic system

Region	Origin of connector fibres	Site of synapse
Sympathetic		
Head and neck	T1–T5	Cervical ganglia
Upper limb	T2–T6	Inferior cervical and 1st thoracic ganglia
Lower limb	T10–L2	Lumbar and sacral ganglia
Heart	T1–T5	Cervical and upper thoracic ganglia
Lungs	T2–T4	Upper thoracic ganglia
Abdominal and pelvic viscera	T6–L2	Coeliac and subsidiary ganglia
Parasympathetic		
Head and neck	Cranial nerves 3, 7, 9, 10	Various parasympathetic macroscopic ganglia
Heart	Cranial nerve 10	Ganglia in vicinity of heart
Lungs	Cranial nerve 10	Ganglia in hila of lungs
Abdominal and pelvic viscera	Cranial nerve 10	Microscopic ganglia in walls of viscera (down to transverse colon)
	S2, 3, 4	Microscopic ganglia in walls of viscera

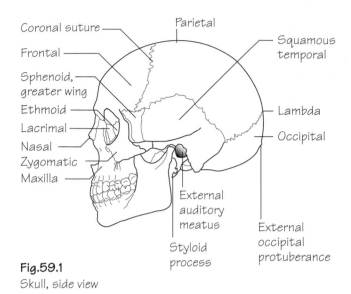

Fig.59.1
Skull, side view

Labels (Fig. 59.1): Coronal suture, Parietal, Squamous temporal, Frontal, Sphenoid, greater wing, Ethmoid, Lacrimal, Nasal, Zygomatic, Maxilla, Lambda, Occipital, External auditory meatus, Styloid process, External occipital protuberance

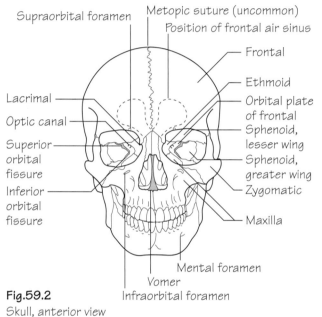

Fig.59.2
Skull, anterior view

Labels (Fig. 59.2): Supraorbital foramen, Metopic suture (uncommon), Position of frontal air sinus, Frontal, Lacrimal, Ethmoid, Optic canal, Orbital plate of frontal, Superior orbital fissure, Sphenoid, lesser wing, Sphenoid, greater wing, Inferior orbital fissure, Zygomatic, Maxilla, Mental foramen, Vomer, Infraorbital foramen

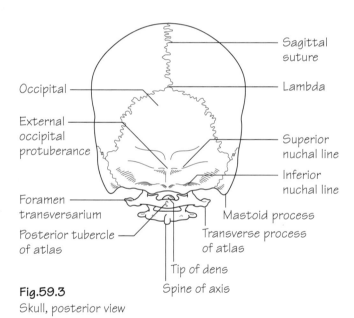

Fig.59.3
Skull, posterior view

Labels (Fig. 59.3): Sagittal suture, Lambda, Occipital, External occipital protuberance, Superior nuchal line, Inferior nuchal line, Foramen transversarium, Posterior tubercle of atlas, Mastoid process, Transverse process of atlas, Tip of dens, Spine of axis

The skull consists of the bones of the *cranium* (making up the *vault* and the *base*) and the bones of the *face*, including the mandible.

The bones of the cranium
The vault of the skull
• The vault of the skull comprises a number of flat bones, each of which consists of two layers of compact bone separated by a layer of cancellous bone (the *diploë*) which contains red bone marrow and a number of *diploic veins*. The bones are the *frontal, parietal, occipital, squamous temporal* and the *greater wing of the sphenoid*. The *frontal air sinuses* are in the frontal bone just above the orbit. The bones are separated by *sutures* which hold the bones firmly together in the mature skull (Figs. 59.1–59.3). Occasionally, the frontal bone may be separated into two halves by a midline *metopic suture*.

• There are a number of *emissary foramina* which transmit *emissary veins*. These establish a communication between the intra- and extracranial veins.

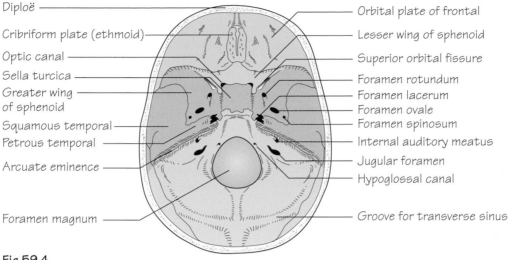

Fig.59.4
The interior of the skull base.
The anterior, middle and posterior cranial fossae are coloured green, red and blue respectively

The interior of the base of the skull

The interior of the base of the skull comprises the *anterior*, *middle* and *posterior cranial fossae* (Fig. 59.4).

The anterior cranial fossa
- **Bones**:
  - Orbital plate of the frontal bone.
  - Lesser wing of the sphenoid.
  - Cribriform plate of the ethmoid.
- **Foramina**:
  - In the cribriform plate (olfactory nerves).
  - Optic canal (optic nerve and ophthalmic artery).
- **Other features**:
  - The orbital plate of the frontal bone forms the roof of the orbit.
  - The *anterior clinoid processes* are lateral to the optic canals.
  - The boundary between the anterior and middle cranial fossae is the sharp posterior edge of the lesser wing of the sphenoid.

The middle cranial fossa
- **Bones**:
  - Greater wing of the sphenoid.
  - Temporal bone.
- **Foramina**:
  - Superior orbital fissure (frontal, lacrimal and nasociliary branches of trigeminal nerve; oculomotor, trochlear and abducent nerves; ophthalmic veins).
  - Foramen rotundum (maxillary branch of trigeminal nerve).
  - Foramen ovale (mandibular branch of trigeminal nerve).
  - Foramen spinosum (middle meningeal artery).
  - Foramen lacerum (internal carotid artery through upper opening (p. xx)).
- **Other features**:
  - The superior orbital fissure is between the greater and lesser wings of the sphenoid.

- In the midline is the *body of the sphenoid* with the *sella turcica* on its upper aspect. It contains the *sphenoidal air sinus*.
- The *foramen lacerum* is the gap between the apex of the petrous temporal bone and the body of the sphenoid, it is covered by cartilage in life.
- The boundary between the middle and posterior cranial fossae is the sharp upper border of the petrous temporal bone.

The posterior cranial fossa
- **Bones**:
  - Petrous temporal (posterior surface).
  - Occipital.
- **Foramina**:
  - Foramen magnum (lower part of medulla, vertebral arteries, spinal accessory nerve).
  - Internal auditory meatus (facial and vestibulocochlear nerves, internal auditory artery).
  - Jugular foramen (glossopharyngeal, vagus and accessory nerves, internal jugular vein).
  - Hypoglossal canal (hypoglossal nerve).
- **Other features**:
  - The jugular foramen is the gap between the occipital and petrous temporal bones.
  - The inner surface of the occipital bone is marked by deep grooves for the transverse and sigmoid venous sinuses. They lead down to the jugular foramen.

Clinical notes

- On an X-ray of the skull, there are markings which may be mistaken for a fracture. These are caused by: (1) the middle meningeal artery; (2) diploic veins; or (3) the sutures, including the infrequent metopic suture.

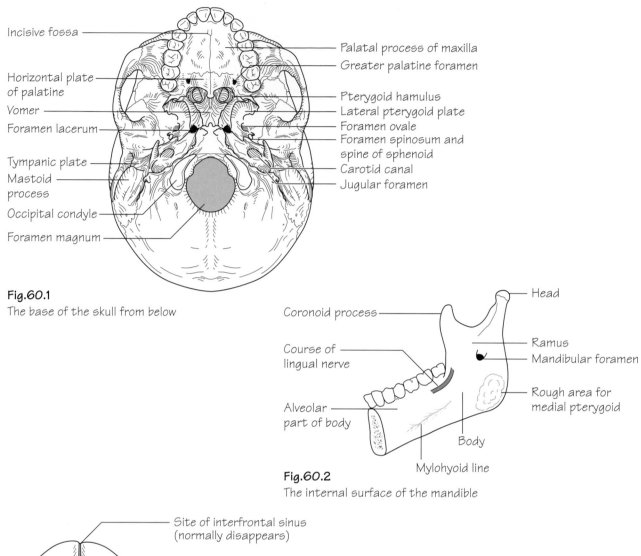

Fig.60.1
The base of the skull from below

Labels (Fig.60.1):
Incisive fossa
Horizontal plate of palatine
Vomer
Foramen lacerum
Tympanic plate
Mastoid process
Occipital condyle
Foramen magnum
Palatal process of maxilla
Greater palatine foramen
Pterygoid hamulus
Lateral pterygoid plate
Foramen ovale
Foramen spinosum and spine of sphenoid
Carotid canal
Jugular foramen

Labels (Fig.60.2):
Coronoid process
Course of lingual nerve
Alveolar part of body
Head
Ramus
Mandibular foramen
Rough area for medial pterygoid
Body
Mylohyoid line

Fig.60.2
The internal surface of the mandible

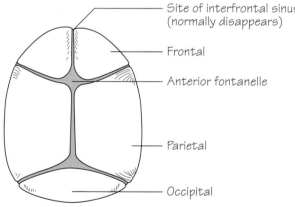

Labels (Fig.60.3):
Site of interfrontal sinus (normally disappears)
Frontal
Anterior fontanelle
Parietal
Occipital

Fig.60.3
A fetal skull, from above

The outside of the base of the skull
(Fig. 60.1)

The anterior part of the cranial base is hidden by the bones of the face. The remainder consists of the bones that were seen in the middle and posterior cranial fossae, but many of the foramina seen on the exterior are not visible inside the cranium.

- **Bones**:
  - Temporal (squamous, petrous and tympanic parts and the styloid process).
  - Sphenoid (body and greater wing). The body bears the medial and lateral pterygoid plates.
- **Foramina**:
  - Foramen magnum (already described).
  - Hypoglossal canal (already described).
  - Stylomastoid foramen (facial nerve).
  - Jugular foramen (already described).
  - Foramen lacerum (the internal carotid through its *internal* opening).
  - Carotid canal (internal carotid artery and sympathetic nerves).
  - Foramen spinosum (already described).
  - Foramen ovale (already described).
- **Other features**:
  - The area between and below the nuchal lines is for the attachment of the extensor muscles of the neck.
  - The *occipital condyles*, for articulation with the atlas, lie on either side of the foramen magnum.
  - The *mastoid process* is part of the petrous temporal bone and contains the *mastoid air cells* (p. xx).
  - The floor of the *external auditory meatus* is formed by the *tympanic plate of the temporal bone*.
  - The carotid canal (see Fig. 64.2) turns inside the temporal bone to run horizontally forwards. It then opens into the posterior wall of the foramen lacerum before turning upwards again to enter the cranial cavity through the internal opening of the foramen.
  - Behind the foramen spinosum is the spine of the sphenoid which lies medial to the mandibular fossa for articulation with the head of the mandible.
  - In front of this is the articular eminence, onto which the head of the mandible moves when the mouth is open.

The bones of the face
(Figs. 59.2 and 60.2)

The bones of the face are suspended below the front of the cranium and comprise the bones of the upper jaw, the bones around the orbit and nasal cavities and the mandible.

- **Bones**:
  - Maxilla.
  - Pterygoid plates of the sphenoid.
  - Palatine.
  - Zygomatic.
  - Nasal.
  - Frontal.
  - Lacrimal.
  - Bones of the orbit and nasal cavities (see below).
- **Foramina**:
  - Supraorbital (supraorbital nerve and vessels).
  - Infraorbital (infraorbital nerve and vessels).
  - Mental (mental nerve and vessels).
  - Greater and lesser palatine foramina (greater and lesser palatine nerves and vessels).
  - Foramina of the incisive fossa (nasopalatine nerves and vessels).
- **Other features**:
  - The pterygoid plates of the sphenoid support the back of the maxilla.
  - Between these two bones is the *pterygomaxillary fissure*, which leads into the *pterygopalatine fossa*.
  - The hard palate is formed by the *palatine process of the maxilla* and the *horizontal plate of the palatine*.
  - The upper teeth are borne in the maxilla.
  - The maxilla contains the large *maxillary air sinus*.
  - The orbital margins are formed by the *frontal*, *zygomatic* and *maxillary* bones.
  - The ethmoid lies between the two orbits and contains the *ethmoidal air cells*.
  - The lacrimal has a fossa for the lacrimal sac.
  - At the back of the orbit are the *greater and lesser wings of the sphenoid* with the *superior orbital fissure* between them. Also, the *optic canal* and the *infraorbital fissure*.
  - The bones of the nasal cavity are the *maxilla*, the *inferior concha*, the *ethmoid*, the *vomer*, the *nasal septum* and the *perpendicular plate of the palatine*.

The **mandible** (Fig. 60.2) consists of the *body* and two *rami*. Each ramus divides into a *coronoid process* and the *head*, for articulation with the mandibular fossa. The *mandibular foramen* transmits the inferior alveolar nerve and vessels.

The development of the skull

- The *base of the skull* ossifies in cartilage and, as the cranial nerves are already present, the bones become ossified around the nerves to form foramina. The last part of the base to ossify is the cartilaginous plate between the body of the sphenoid and the basi-occiput, which is an important growth centre. These two bones finally fuse at 25 years of age.
- The *vault of the skull* ossifies in membrane to form the frontal, parietal, occipital and squamous temporal bones. In the fetus, there are two frontal bones and the diamond-shaped area between these two and the two parietal bones is the *anterior fontanelle* (Fig. 60.3). This closes between 18 months and 2 years of age. There are smaller fontanelles between other bones. Normally, the suture between the two frontal bones disappears to leave a single frontal bone, but occasionally the interfrontal suture will persist so that there is a vertical suture in the midline – the *metopic suture* – which is visible on an X-ray.

Clinical notes

- The neonatal skull is still only partly ossified, so that moulding of the head can occur during birth, the bones sometimes riding over each other. If there is any blockage to the drainage of cerebrospinal fluid, the skull will expand (*hydrocephalus*) with bulging of the anterior fontanelle. This condition is usually associated with other anomalies of the nervous system.

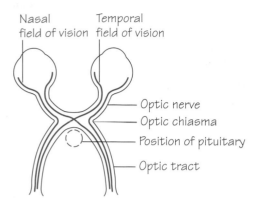

Fig.61.1
The optic chiasma.
Only the fibres from the nasal side of the retina
(i.e. the temporal fields of vision) cross in the chiasma

Labels: Nasal field of vision; Temporal field of vision; Optic nerve; Optic chiasma; Position of pituitary; Optic tract

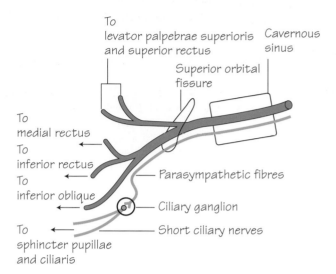

Fig.61.2
The oculomotor nerve.
Parasympathetic fibres are shown in orange

Labels: To levator palpebrae superioris and superior rectus; Cavernous sinus; Superior orbital fissure; To medial rectus; To inferior rectus; To inferior oblique; Parasympathetic fibres; Ciliary ganglion; Short ciliary nerves; To sphincter pupillae and ciliaris

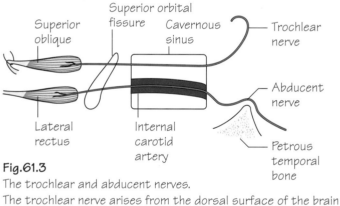

Fig.61.3
The trochlear and abducent nerves.
The trochlear nerve arises from the dorsal surface of the brain

Labels: Superior oblique; Superior orbital fissure; Cavernous sinus; Trochlear nerve; Abducent nerve; Petrous temporal bone; Lateral rectus; Internal carotid artery

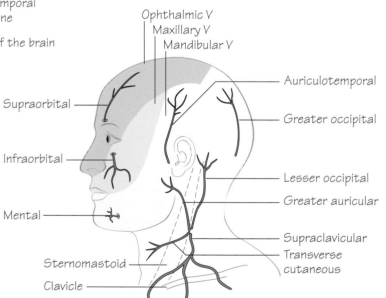

Fig.61.4
The main sensory nerves of the face and neck

Labels: Ophthalmic V; Maxillary V; Mandibular V; Auriculotemporal; Greater occipital; Supraorbital; Infraorbital; Lesser occipital; Greater auricular; Mental; Supraclavicular; Transverse cutaneous; Sternomastoid; Clavicle

The head and neck are supplied by the first four spinal and the 12 cranial nerves.

The spinal nerves

- **C1:** supplies the small suboccipital muscles. Its anterior ramus joins the hypoglossal nerve but leaves it later to form the *descendens hypoglossi*.
- **C2:** the posterior ramus forms the *greater occipital nerve*, which is sensory to the scalp.
- **The posterior rami of C2, C3 and C4** provide muscular and sensory branches to the back. Their anterior rami provide muscular branches, including the *descendens cervicalis* (see hypoglossal nerve, p. 145). They also supply sensory branches: the *greater auricular, lesser occipital, anterior cutaneous* and the *three supraclavicular nerves* (Fig. 61.4). The greater auricular supplies the skin in the parotid region, the only sensory supply to the face which is not derived from the trigeminal. The others supply the skin of the neck and the upper part of the thorax.
- **The remaining cervical nerves (C5–C8)** join the brachial plexus.

The cranial nerves (Figs. 61.1, 61.2, 61.3, and 61.4)

- **I. The olfactory nerve:** the cell bodies of the olfactory nerve are in the nasal mucosa. Their axons form the olfactory nerves which ascend through the cribriform plate to synapse in the olfactory bulb of the brain.
- **II. The optic nerve (Fig. 61.1):** the eye and optic nerve develop as an outgrowth of the embryonic brain and the nerve is, therefore, enveloped in meninges. The cell bodies are in the retina and the axons pass back in the optic nerve to the *optic chiasma*, where the axons from the nasal halves of the retina cross over but those from the temporal side continue on the same side. They, then, form the *optic tract* on each side.
- **III. The oculomotor nerve (Fig. 61.2):** arises from the brain just in front of the pons, traverses the cavernous sinus and enters the orbit through the *superior orbital fissure*. It supplies the levator palpebrae superioris, superior, inferior and medial rectus muscles and the inferior oblique. It also carries parasympathetic fibres to the *ciliary ganglion* where the fibres synapse and then pass in the short ciliary nerves to the sphincter pupillae and the ciliary muscles (see Chapter 74).
- **IV. The trochlear nerve (Fig. 61.3):** arises from the dorsal surface of the brain just behind the *inferior colliculus*, winds round the midbrain and enters the *cavernous sinus*. It enters the orbit through the *superior orbital fissure* and supplies the superior oblique.

Clinical notes

- **The optic nerve and chiasma**: the optic nerve is really part of the brain and, like the brain and spinal cord, is unable to regenerate if it is divided. For this reason, too, transplantation of the eye is, at present, impossible. In contrast, the cornea contains no blood vessels or lymphatics and can, therefore, be transplanted very effectively without fear of rejection.

  Division of one optic tract will lead to loss of vision on the temporal side of the retina of the same side and the nasal side of the retina of the opposite side (Fig. 61.1). There will thus be loss of one side of the total field of vision (*homonymous hemianopia*).

  A lesion that affects the chiasma itself, such as a tumour of the pituitary, will affect the fibres that cross over and will, therefore, cause loss of the temporal fields of vision in both eyes (*bitemporal hemianopia*).

- **Papilloedema**: as the subarachnoid space continues around the optic nerve as far as the back of the eyeball, any rise in intracranial pressure, such as occurs with intracranial tumours, can cause swelling of the optic disc (*papilloedema*) which can be seen with an ophthalmoscope.

- **The oculomotor nerve**: interruption of the oculomotor nerve will cause paralysis of all of the extrinsic muscles of the eye, except for the lateral rectus and the superior oblique. There will, thus, be a divergent squint owing to the unopposed action of these two muscles, and this will give rise to double vision (*diplopia*). This, however, will only occur when the eyelid is held up, as paralysis of the levator palpebrae superioris leads to drooping of the upper eyelid (*ptosis*). There will also be dilatation of the pupil owing to loss of the parasympathetic fibres carried in the oculomotor nerve.

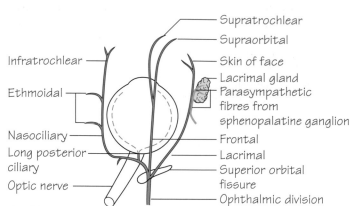

Supratrochlear
Supraorbital
Infratrochlear
Skin of face
Lacrimal gland
Ethmoidal
Parasympathetic fibres from sphenopalatine ganglion
Nasociliary
Frontal
Long posterior ciliary
Lacrimal
Superior orbital fissure
Optic nerve
Ophthalmic division

Fig.62.1
The course and branches of the ophthalmic division of the trigeminal nerve.
Parasympathetic fibres are shown in orange

Lacrimal gland
Ophthalmic division
Foramen rotundum
Maxillary division
Mandibular division
Sphenopalatine ganglion
Infraorbital
Nasal branches
Greater palatine
Lesser palatine
Sphenopalatine
Posterior superior dental
Incisive fossa

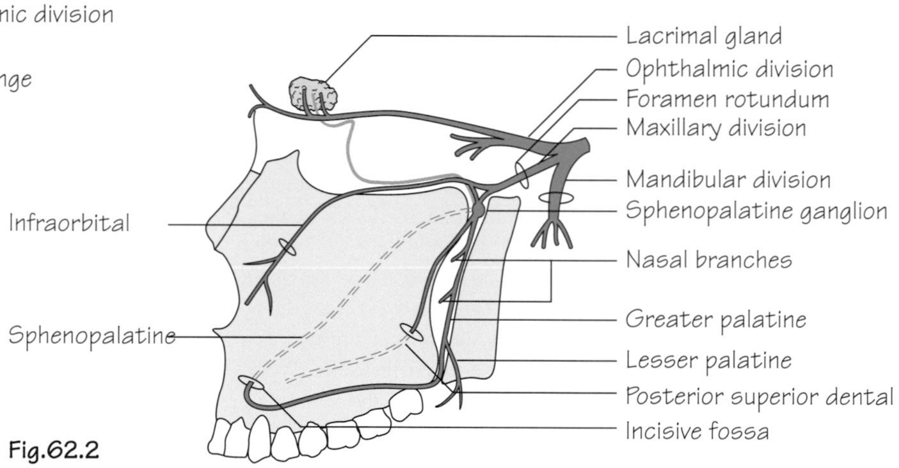

Fig.62.2
The course and branches of the maxillary division of the trigeminal nerve.
Parasympathetic fibres are shown in orange

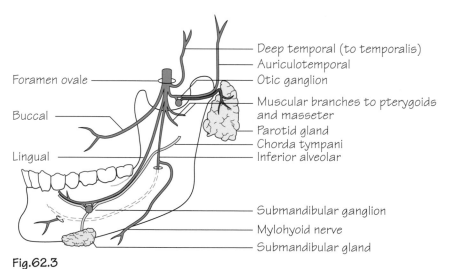

Deep temporal (to temporalis)
Auriculotemporal
Foramen ovale
Otic ganglion
Buccal
Muscular branches to pterygoids and masseter
Parotid gland
Chorda tympani
Lingual
Inferior alveolar
Submandibular ganglion
Mylohyoid nerve
Submandibular gland

Fig.62.3
The course and main branches of the mandibular division of the trigeminal nerve.
The fibres of the chorda tympani are shown in yellow

The trigeminal nerve (Figs. 61.4, 62.1, 62.2, and 62.3) arises from the brain at the side of the pons by a motor and a sensory root. The sensory root carries the *trigeminal ganglion*, which consists of the cell bodies of the sensory axons and lies in a depression on the petrous temporal bone. It then divides into *ophthalmic*, *maxillary* and *mandibular* divisions. The motor root forms part of the mandibular division.

The ophthalmic division (Fig. 62.1)

This traverses the cavernous sinus and enters the orbit via the *superior orbital fissure* where it divides into *frontal*, *lacrimal* and *nasociliary* branches. The *frontal nerve* lies just under the roof of the orbit and divides into *supraorbital* and *supratrochlear nerves*, which emerge from the orbit and supply the front of the scalp. The *lacrimal nerve* lies laterally and supplies the skin of the eyelids and face. It also carries parasympathetic secretomotor fibres from the *sphenopalatine ganglion* to the *lacrimal gland*. The *nasociliary nerve* crosses the optic nerve and runs along the medial wall of the orbit to emerge onto the face as the *infratrochlear nerve*. It gives off the *ethmoidal nerves* to the ethmoidal sinuses and the *long ciliary nerves* to the eye which carry sensory fibres from the cornea and sympathetic fibres to the dilator pupillae. *All branches of the ophthalmic division are sensory.*

The maxillary division (Fig. 62.2)

This leaves the cranial cavity through the foramen rotundum and enters the *pterygopalatine fossa*. It has the *sphenopalatine ganglion* attached to it, which transmits parasympathetic fibres to the lacrimal gland via communications with the lacrimal nerve. The branches of the maxillary nerve are the *greater* and *lesser palatine nerves* to the hard and soft palates and the *sphenopalatine nerve* to the nasal cavity and thence, via the nasal septum, to the incisive fossa to supply the hard palate. The *posterior superior dental nerve* enters the back of the maxilla and supplies the teeth. The maxillary nerve leaves the sphenopalatine fossa via the inferior orbital fissure, travels in the floor of the orbit, where it gives the *middle* and *anterior superior dental nerves*, and emerges onto the face through the infraorbital foramen as the *infraorbital nerve*. *All branches of the maxillary division are sensory.*

The mandibular division (Fig. 62.3)

This leaves the cranial cavity through the foramen ovale and immediately breaks up into branches. The mainly sensory *inferior alveolar nerve* enters the mandibular foramen to supply the teeth before emerging onto the face as the *mental nerve*. This nerve does have one motor branch, the *mylohyoid nerve*, which supplies the mylohyoid and the anterior belly of the digastric. The *lingual nerve* lies close to the mandible just behind the third molar and then passes forwards to supply the tongue. It is joined by the *chorda tympani* which carries taste fibres from the anterior two-thirds of the tongue and parasympathetic secretomotor fibres to the submandibular and sublingual salivary glands. These synapse in the submandibular ganglion which is attached to the lingual nerve. The *auriculotemporal nerve* supplies sensory fibres to the side of the scalp. It also carries parasympathetic secretomotor fibres, which have synapsed in the otic ganglion, to the parotid gland. The *buccal nerve* carries sensory fibres from the face. There are *muscular branches* to the muscles of mastication, including the *deep temporal nerves* which supply temporalis. *The mandibular division, thus, contains both motor and sensory branches.*

Clinical notes

- **Trigeminal neuralgia**: this is a condition, most common in patients over the age of 50 years, in which there is intermittent severe pain in one or more divisions of the trigeminal nerve. The pain is intense and may make the patient screw up his or her face but, fortunately, it does not usually occur at night. Drug treatment is available but, when this is ineffectual, it may be necessary to destroy all or part of the trigeminal ganglion by thermocoagulation. Otherwise, the sensory root of the trigeminal may be sectioned selectively by intracranial surgery.
- **Trigeminal herpes**: like the spinal nerves, the trigeminal nerve may be affected by *herpes zoster*, a virus infection that causes a painful rash in the distribution of the nerve (see also Chapter 78). The most serious form of this is when it affects the ophthalmic division as the rash may then occur on the surface of the eye with the possibility of corneal scarring.
- **Dental anaesthesia**: the maxillary and mandibular divisions are of particular importance to dentists, and techniques are available for blocking all the nerves that supply the teeth. This is normally performed on the side of the lesion but, in the case of the incisor teeth, it is necessary to block both sides as the nerves of each side cross the midline to supply the incisors of the opposite side.

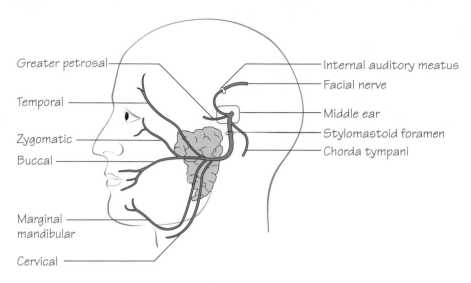

Fig.63.1
The course of the facial nerve.
The nerve passes through the middle ear and the parotid gland

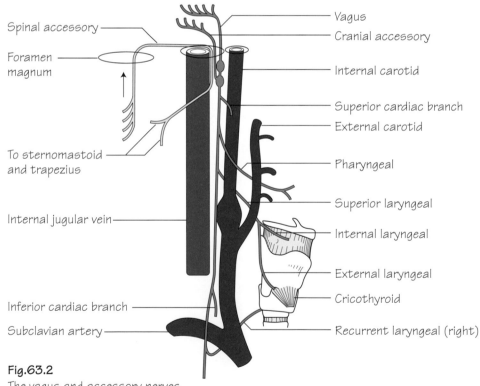

Fig.63.2
The vagus and accessory nerves.
The spinal root of the accessory is shown in yellow

- **VI. The abducent nerve (see Fig. 61.3)**: leaves the brain at the posterior border of the pons and has a long intracranial course to the cavernous sinus, where it is closely applied to the internal carotid artery, and thence to the orbit via the superior orbital fissure. It supplies the lateral rectus.

- **VII. The facial nerve (Fig. 63.1)**: this leaves the brain near the cerebellum and passes laterally into the internal auditory meatus. It reaches the medial wall of the middle ear and turns backwards and downwards to leave the skull via the stylomastoid foramen. It then traverses the parotid gland, in which it divides into five branches (*temporal*, *zygomatic*, *buccal*, *marginal mandibular* and *cervical*), which are distributed to the muscles of facial expression, the platysma and the posterior belly of the digastric. In the middle ear, it gives off the *greater petrosal branch* which carries parasympathetic fibres to the sphenopalatine ganglion and thence to the lacrimal gland. In the middle ear, it also gives off the *chorda tympani* which joins the lingual nerve and is distributed with it. Sensory fibres in the chorda tympani have their cell bodies in the *geniculate ganglion* which lies on the facial nerve where it turns downwards.

- **VIII. The vestibulocochlear (auditory) nerve**: this leaves the brain next to the facial nerve and enters the internal auditory meatus. It divides into *vestibular* and *cochlear* nerves.

- **IX. The glossopharyngeal nerve (Fig. 70.1)**: leaves the brain at the side of the medulla and passes through the jugular foramen. It then curves forwards between the internal and external carotid arteries to enter the pharynx between the superior and middle constrictors. It supplies sensory fibres to the posterior one-third of the tongue (including taste) and the pharynx. It also gives a branch to the carotid body and sinus.

- **X. The vagus nerve (Fig. 63.2)**: arises from the side of the medulla and passes through the jugular foramen. It is joined by the accessory nerve, but the spinal root of the accessory leaves it again almost immediately. The cranial root is distributed with the vagus (hence the name) and is accessory to it. The vagus carries two ganglia for the cell bodies of its sensory fibres. It descends between the internal carotid artery and the jugular vein, within the carotid sheath, and enters the thorax where its further course is described in Chapters 4 and 5. In the neck, the vagus and cranial root of the accessory give the following branches:
  - The *pharyngeal branch* which runs below and parallel to the glossopharyngeal nerve and supplies the striated muscle of the palate and pharynx.
  - *Superior* and *inferior cardiac branches* which descend into the thorax to take part in the cardiac plexuses.
  - The *superior laryngeal nerve* which divides into *internal* and *external laryngeal nerves*. The former enters the larynx by piercing the thyrohyoid membrane and is sensory to the larynx above the level of the vocal cords, and the latter is motor to the cricothyroid muscle.
  - The *recurrent laryngeal nerve*. On the right side, it loops under the subclavian artery before ascending to the larynx behind the common carotid artery. On the left side, it arises from the vagus just below the arch of the aorta and ascends to the larynx in the groove between the trachea and oesophagus. The recurrent laryngeal nerves supply all the muscles of the larynx except for cricopharyngeus and are sensory to the larynx below the vocal cords.

- **XI. The accessory nerve (Fig. 63.2)**: the *cranial root* arises from the side of the medulla with the vagus and is distributed with it. The *spinal root* arises from the side of the upper five segments of the spinal cord, enters the cranial cavity through the foramen magnum and joins the vagus. It leaves the vagus below the jugular foramen and passes backwards to enter sternomastoid, which it supplies. It then crosses the posterior triangle to supply trapezius (see Fig. 66.3).

- **XII. The hypoglossal nerve (Figs. 64.1 and 70.1)**: arises from the side of the medulla ventral to the vagus and cranial accessory and passes through the hypoglossal canal. Below, the skull it is joined by the anterior ramus of C1 and it then runs downwards and forwards, across the carotid sheath and the upward loop of the lingual artery, to enter the tongue. It supplies the intrinsic and extrinsic muscles of the tongue. It gives off the *descendens hypoglossi*, but this is actually composed of fibres from *C1*. This joins the *descendens cervicalis*, derived from C2 and C3, to form the *ansa cervicalis*. From this, branches arise to supply the 'strap muscles', i.e. sternothyroid, sternohyoid, thyrohyoid and omohyoid.

Clinical notes

- **Abducent nerve lesions**: the abducent nerve has a very long intracranial course and so is often affected in intracranial conditions, such as raised intracranial pressure, and aneurysms of the internal carotid artery where it traverses the cavernous sinus. The effect is a convergent squint, diplopia – worst when looking to the affected side – and inability to abduct the eye (i.e. to direct the affected eye laterally).

- **Facial nerve lesions**: the facial nerve may be affected in its intracranial course, in the stylomastoid canal or on the face. The result will be paralysis of the facial musculature and, in the case of intracranial conditions, loss of taste in the anterior two-thirds of the tongue. In upper motor neurone lesions, i.e. those affecting the pathway from the cerebral cortex to the facial nerve nucleus in the pons, the upper part of the face is less affected than the lower because this part of the face is bilaterally represented in the brain. As there is no loss of tone in the muscles in this condition, the unsightly drooping and asymmetry of the face do not occur. The effects of paralysis of the muscles are described in Chapter 72.

- **Bell's palsy**: this is a condition of unknown aetiology in which the facial nerve is affected in the stylomastoid canal. It leads to unilateral facial paralysis. Most cases resolve spontaneously, but others may require treatment.

- **Parotid tumours**: malignant tumours of the parotid often affect the facial nerve as it passes through the gland. The nerve is also vulnerable during surgery on the gland.

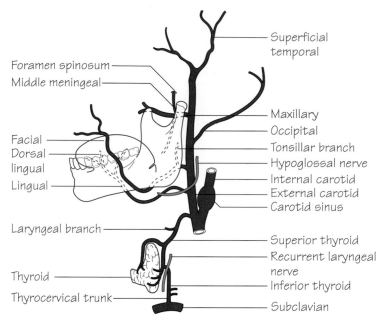

Foramen spinosum
Middle meningeal
Facial
Dorsal lingual
Lingual
Laryngeal branch
Thyroid
Thyrocervical trunk

Superficial temporal
Maxillary
Occipital
Tonsillar branch
Hypoglossal nerve
Internal carotid
External carotid
Carotid sinus
Superior thyroid
Recurrent laryngeal nerve
Inferior thyroid
Subclavian

Fig.64.1
The course and main branches of the external carotid artery.
The inferior thyroid artery is also shown

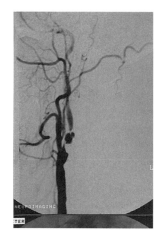

Fig.64.3
Carotid angiogram showing internal carotid stenosis

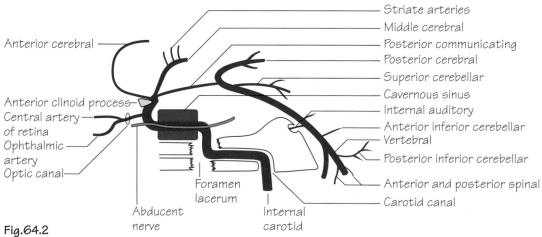

Anterior cerebral
Anterior clinoid process
Central artery of retina
Ophthalmic artery
Optic canal
Abducent nerve
Foramen lacerum
Internal carotid

Striate arteries
Middle cerebral
Posterior communicating
Posterior cerebral
Superior cerebellar
Cavernous sinus
Internal auditory
Anterior inferior cerebellar
Vertebral
Posterior inferior cerebellar
Anterior and posterior spinal
Carotid canal

Fig.64.2
The intracranial course of the internal carotid artery.
The intracranial parts of the two vertebral arteries are also shown diagrammatically
although they are in a different plane

The common carotid artery

Arises from the brachiocephalic artery on the right and from the arch of the aorta on the left (Chapter 6). Each common carotid passes up the neck in the carotid sheath (Fig. 66.1), along with the internal jugular vein and the vagus nerve. At the level of the upper border of the thyroid cartilage it divides into internal and external carotid arteries. There are no branches.

The external carotid artery (Fig. 64.1)

Ascends in the neck a little in front of the internal carotid to divide into its two terminal branches, the *maxillary* and *superficial temporal arteries*, in the substance of the parotid gland.

Branches:

• **The superior thyroid artery**: runs downwards on the side of the pharynx before passing forwards to the upper pole of the thyroid gland where it divides into two branches. The upper branch follows the upper border of the gland towards the isthmus and the lower passes down the posterior border to anastomose with the inferior thyroid artery. There are a number of branches to the larynx.

• **The lingual artery**: arises at the level of the tip of the greater horn of the hyoid and loops upwards for a short distance before running forwards deep to hyoglossus to enter and supply the tongue. It gives a number of *dorsal lingual arteries*. The upward loop of the lingual is crossed by the hypoglossal nerve.

• **The facial artery**: travels forwards, deep to the mandible, where it is embedded in the back of the submandibular gland. It then curls round the lower border of the mandible to reach the face. Here, it follows a tortuous course at the side of the mouth and lateral to the nose to reach the medial angle of the eye where it anastomoses with branches of the ophthalmic artery. It gives off a *tonsillar branch* in the neck, *superior* and *inferior labial branches* and *nasal branches*. The facial arteries anastomose very freely across the midline and with other arteries on the face.

• **The occipital artery**: passes backwards, medial to the mastoid process, and supplies the back of the scalp.

• **The superficial temporal artery**: emerges from the parotid gland and runs up in front of the ear where its pulsations may be felt. It is distributed to the side of the scalp and the forehead.

• **The posterior auricular artery**: supplies the scalp and the pinna of ear. One of its branches enters the middle ear to supply the facial nerve.

• **The ascending pharyngeal artery**: arises from the posterior aspect of the external carotid and runs up the lateral wall of the pharynx, which it supplies. It also supplies the soft palate and sends branches to the meninges

• **The maxillary artery**: emerges from the parotid gland and passes deep to the neck of the mandible. It ends by entering the pterygopalatine fossa through the pterygomaxillary fissure. Its principal branches are to the local muscles including the *deep temporal arteries* to temporalis and:

  • The *inferior alveolar artery*: enters the mandibular canal to supply the teeth.

  • The *middle meningeal artery*: runs upwards to pass through the foramen spinosum. Inside the skull, it passes laterally and then ascends on the squamous temporal bone in a deep groove, which it shares with the corresponding vein. The *anterior* branch passes upwards and backwards towards the vertex and the *posterior* branch passes backwards. It supplies the dura mater and the bones of the cranium.

  • *Branches which accompany the branches of the maxillary nerve* in the pterygopalatine fossa have the same names.

The internal carotid artery (Figs. 64.2, 64.3 and 65.4)

At its origin from the common carotid artery, it is enlarged to form the *carotid sinus*, a slight dilatation which has baroreceptors supplied by the glossopharyngeal nerve in its wall. Associated with this is the *carotid body*, a *chemoreceptor* supplied by the same nerve. The internal carotid has no branches in the neck. It enters the cranial cavity via the carotid canal in the petrous temporal bone, accompanied by a sympathetic plexus. Within the skull it passes forwards in the cavernous sinus and then turns backwards behind the anterior clinoid process to break up into its three terminal branches.

Branches:

• **The ophthalmic artery**: enters the orbit through the superior orbital fissure and follows the nasociliary nerve. It gives the important *central retinal artery*, which enters the optic nerve and supplies the retina. Other branches are described on p. 169.

• **The anterior cerebral artery**: winds round the *genu of the corpus callosum* and supplies the front and medial surfaces of the cerebral hemisphere. It anastomoses with its fellow of the opposite side to form the anterior component of the circle of Willis.

• **The middle cerebral artery**: traverses the *lateral sulcus* on the lateral surface of the hemisphere and supplies the hemisphere (including the main motor and sensory areas), as well as giving the *striate arteries* which supply deep structures including the *internal capsule*.

• **The posterior communicating artery**: a small artery which passes backwards to join the *posterior cerebral artery* (a terminal branch of the vertebral artery).

These arteries and the communications between them form the *circle of Willis*, so that there is (usually) free communication between the branches of the two internal carotid arteries across the midline. There is, however, considerable variation in the arrangement of the circle.

Clinical notes

• **Extradural haemorrhage**: the middle meningeal artery is at risk in head injuries and may bleed to produce a collection of blood (haematoma) under the dura. The patient may apparently recover from the initial concussion and then, after a symptomless period, begin to suffer the effects of increased intracranial pressure. It is, therefore, important to keep a patient who has been unconscious under close observation for some time after the injury.

• **Central retinal artery occlusion**: the central artery of the retina is the most important branch of the ophthalmic artery as it is an *end-artery*, i.e. it does not anastomose with any other arteries so that a collateral circulation cannot occur. Occlusion will, therefore, cause immediate blindness.

• **Intracranial (berry) aneurysms**: the intracranial arteries have much thinner walls than arteries elsewhere, and in the region of side branches there are sometimes defects in the media. It is at these sites that aneurysms can occur, particularly with advancing age. They may rupture, producing a subarachnoid haemorrhage, which may be immediately fatal or may need surgical treatment.

• **Occlusion of the internal carotid artery**: the internal carotid artery may be narrowed (*stenosis*) or occluded by atheroma (Fig. 64.3) but, if the circle of Willis is intact, the artery of the other side may take over the territory of the occluded artery.

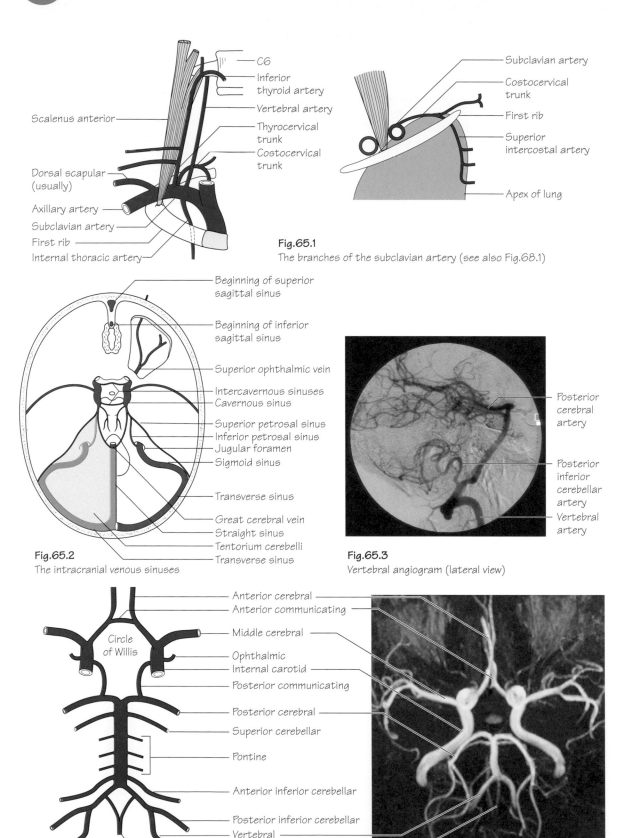

Fig.65.1
The branches of the subclavian artery (see also Fig.68.1)

Fig.65.2
The intracranial venous sinuses

Fig.65.3
Vertebral angiogram (lateral view)

Fig.65.4 (see Ch. 64)
Diagram and magnetic resonance angiogram (MRA) showing the distribution
of the internal carotid and vertebral arteries within the cranial cavity (see p 147)

The subclavian artery (Figs. 65.1 and 68.1)

The subclavian artery arises from the *brachiocephalic artery* on the right and the *arch of the aorta* on the left. It arches across the upper surface of the 1st rib to become the *axillary artery*. It is in close contact with the apex of the lung and lies behind scalenus anterior at the root of the neck.

Branches:
- **The internal thoracic artery:** see p. 21.
- **The vertebral artery:** runs upwards to enter the foramen transversarium of the 6th cervical vertebra. It passes through corresponding foramina in the other cervical vertebra to reach the upper surface of the atlas. Here, it turns medially in a groove and then enters the cranial cavity through the *foramen magnum*. Here, it joins its fellow of the opposite side to form the *basilar artery*. It gives off the *anterior* and *posterior spinal arteries* which descend to supply the spinal cord, and the *posterior inferior cerebellar artery* which supplies not only the cerebellum but also the medulla. The basilar artery passes forwards on the anterior surface of the medulla and pons and gives the *anterior inferior cerebellar artery*, branches to the brainstem and to the inner ear (the *internal auditory artery*) and ends by dividing into the *superior cerebellar* and *posterior cerebral arteries*. The latter is joined by the *posterior communicating artery* (p. 147) to form the posterior part of the circle of Willis (see Chapter 64).
- **The costocervical trunk:** a small artery that passes backwards to supply the muscles of the back. It also supplies the *superior intercostal artery* (see Chapter 6).
- **The thyrocervical trunk:** gives off the *superficial cervical* and *suprascapular arteries* and then passes medially as the *inferior thyroid artery* across the vertebral artery to reach the middle of the posterior border of the thyroid. It has a variable relation to the recurrent laryngeal nerve, lying in front or behind it, but may branch early with the nerve passing between the branches.
- **The dorsal scapular artery:** usually given off from the third part (Fig. 68.1) but may arise in common with the superficial cervical artery. It descends along the medial border of the scapula.

The veins

The veins of the brain drain into *dural venous sinuses* (Fig. 65.2). The most important of these are:
- **The superior sagittal sinus:** passes backwards in the midline in the attached border of the falx cerebri from just above the cribriform plate to the occipital region, where it communicates with the *straight sinus*, and then turns to the right to form the *right transverse sinus*. It then winds down on the back of the petrous temporal as the *sigmoid sinus*, which passes through the right jugular foramen to form the *right internal jugular vein*.
- **The inferior sagittal sinus:** begins near the origin of the superior sagittal sinus and runs in the free border of the falx cerebri. It is joined by the great cerebral vein to form the *straight sinus* which lies in the attachment of the falx to the tentorium cerebelli. The straight sinus turns to the left to form the left transverse sinus and then the sigmoid sinus. The latter leaves the skull through the left jugular foramen.
- **The cavernous sinus:** this lies at the side of the pituitary fossa and contains the internal carotid artery. It receives the superior and inferior ophthalmic veins and is connected to some smaller sinuses – the *superior* and *inferior petrosal sinuses* and the *sphenoidal sinus*. The two cavernous sinuses are joined in front and behind the pituitary by the *intercavernous sinuses*.
- **The emissary veins:** see p. 136.
- **The internal jugular vein:** passes down the neck from the jugular foramen in the carotid sheath along with the internal and common carotid arteries and the vagus nerve. It ends by joining the subclavian vein to form the *brachiocephalic vein*. It receives veins corresponding to the branches of the external carotid artery (*facial, lingual, pharyngeal* and the *superior* and *middle thyroid veins*). The *inferior thyroid veins* pass downwards in front of the trachea to open into the left brachiocephalic vein.
- **The external jugular vein:** begins in the parotid gland by the joining of the *retromandibular vein* with other small veins. It passes obliquely across sternomastoid to open into the subclavian vein. It receives the *transverse cervical, suprascapular* and *anterior jugular veins* near its lower end (Fig. 66.3).
- **The anterior jugular vein:** begins below the chin and runs down the neck near the midline. It then passes deep to sternomastoid to join the external jugular vein.
- **The subclavian vein:** lies in a groove on the 1st rib but is separated from the subclavian artery by the scalenus anterior. It receives the external jugular vein, veins corresponding to the branches of the subclavian artery and, at its junction with the internal jugular vein, the *thoracic duct* on the left and the *right lymph duct* on the right.
- **The vertebral vein:** this is formed at the level of the 6th foramen transversarium from the vertebral plexus of veins that accompany the vertebral artery.

Clinical notes

- **Aberrant subclavian artery**: in about 2% of the population, part of the embryonic aortic arch system fails to disappear and the right subclavian artery is the last branch of the aortic arch. It passes behind the oesophagus in the thorax and then ascends to reach its usual position on the 1st rib. Such an artery (*arteria lusoria*) may cause *dysphagia* (difficulty in swallowing).
- **Cavernous sinus thrombosis**: the facial vein communicates around the orbit with tributaries of the ophthalmic veins, so that infections of the face may spread to the cavernous sinus and cause thrombosis if not properly treated.

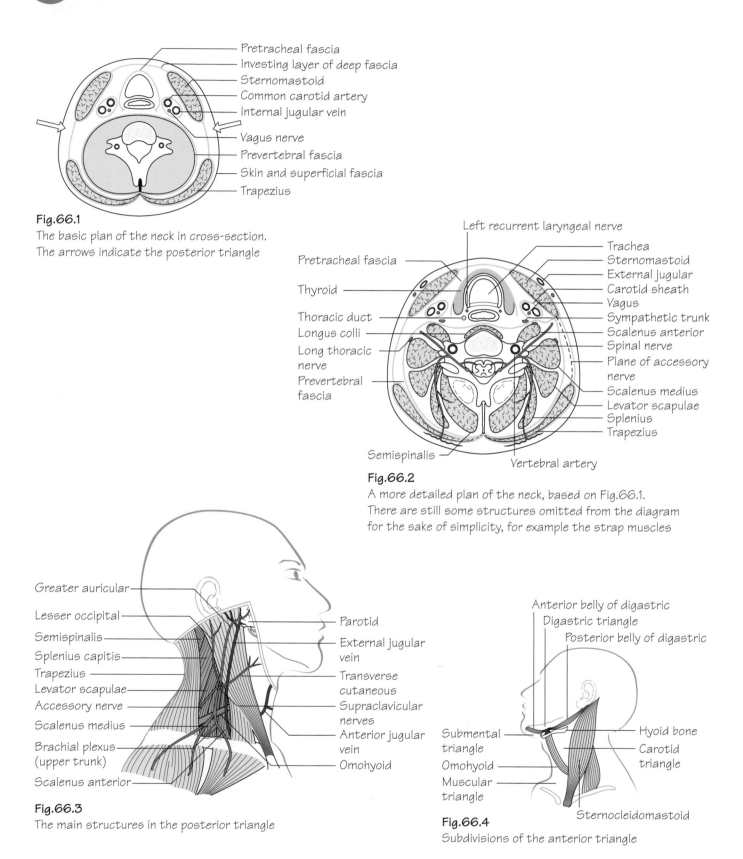

Fig.66.1
The basic plan of the neck in cross-section. The arrows indicate the posterior triangle

Pretracheal fascia
Investing layer of deep fascia
Sternomastoid
Common carotid artery
Internal jugular vein
Vagus nerve
Prevertebral fascia
Skin and superficial fascia
Trapezius

Fig.66.2
A more detailed plan of the neck, based on Fig.66.1. There are still some structures omitted from the diagram for the sake of simplicity, for example the strap muscles

Left recurrent laryngeal nerve
Trachea
Sternomastoid
External jugular
Carotid sheath
Vagus
Sympathetic trunk
Scalenus anterior
Spinal nerve
Plane of accessory nerve
Scalenus medius
Levator scapulae
Splenius
Trapezius
Vertebral artery
Semispinalis

Pretracheal fascia
Thyroid
Thoracic duct
Longus colli
Long thoracic nerve
Prevertebral fascia

Fig.66.3
The main structures in the posterior triangle

Greater auricular
Lesser occipital
Semispinalis
Splenius capitis
Trapezius
Levator scapulae
Accessory nerve
Scalenus medius
Brachial plexus (upper trunk)
Scalenus anterior

Parotid
External jugular vein
Transverse cutaneous
Supraclavicular nerves
Anterior jugular vein
Omohyoid

Fig.66.4
Subdivisions of the anterior triangle

Anterior belly of digastric
Digastric triangle
Posterior belly of digastric
Submental triangle
Omohyoid
Muscular triangle
Hyoid bone
Carotid triangle
Sternocleidomastoid

The neck consists essentially of five blocks of tissue running longitudinally (Figs. 66.1 and 66.2). These are as follows:

1. **The cervical vertebrae** surrounded by a number of muscles and enclosed in a dense layer of *prevertebral fascia*.
2. **The pharynx and larynx**, partially enclosed in a thin layer of *pretracheal fascia*. Below the level of C6, these give way to the oesophagus and trachea.
3 and 4. **Two vascular packets** consisting of the common and internal carotid arteries, the internal jugular vein and the vagus nerve, all enclosed in the fascial *carotid sheath*.
5. **An outer enclosing sheath** consisting of the sternomastoid and trapezius and the *investing layer of deep fascia of the neck*.

The anterior triangle

The *anterior triangle* (Fig. 66.4) is bounded by:
- The lower border of the mandible and its backward continuation.
- The anterior border of sternomastoid.
- The midline of the neck.

The anterior triangle is subdivided into:

1 The digastric triangle, bounded by:
- The lower border of the mandible.
- The two bellies of the digastric.

2 The *carotid triangle*, bounded by:
- The superior belly of the omohyoid.
- The posterior belly of the digastric.
- The anterior border of sternomastoid.

3 The *muscular triangle*, bounded by:
- The superior belly of the omohyoid.
- The anterior border of sternomastoid.
- The midline of the neck.

The contents of these triangles are mostly structures that are continuous, without interruption, from one triangle to another, so that it is more convenient to describe them individually in other chapters.

The posterior triangle (Fig. 66.3)

The *posterior triangle* is bounded by:
- The posterior border of *sternomastoid*.
- The anterior border of *trapezius*.
- The middle part of the clavicle.

Stretching between the two muscles is the *investing layer of deep fascia* which splits to enclose them and continues to the anterior triangle. Embedded in the deep fascia is the *spinal part of the accessory nerve* which leaves the sternomastoid about halfway down its posterior border and passes into trapezius two fingers' breadth above the clavicle. It supplies both muscles. Four cutaneous nerves (*transverse cervical, supraclavicular, greater auricular* and *lesser occipital*) also emerge near the accessory nerve and supply the skin of the neck and the upper part of the chest. The external jugular vein begins near the upper end of sternomastoid and runs down obliquely across this muscle to enter the subclavian vein. It is joined by the anterior jugular and other small veins at its lower end. The inferior belly of the omohyoid muscle crosses the lower part of the triangle.

The floor of the posterior triangle is the *prevertebral fascia*, deep to which lie, from below upwards: scalenus anterior, scalenus medius and posterior, levator scapulae and splenius capitis.

Structures deep to the prevertebral fascia

- **The upper, middle and lower trunks of the brachial plexus** which emerge between the scalenus anterior and the scalenus medius, the lower trunk resting on the 1st rib.
- **The supraclavicular branches** of the brachial plexus (p. 87).

Clinical notes

- **The external jugular vein**: this vein is easily visible on the surface of sternomastoid when the patient is lying down. There are no valves between this vein and the heart, so that if the pressure is raised in the right side of the heart, the external jugular will be engorged even when the patient is sitting up.
- **Air embolism**: if one of the neck veins is opened, with the patient in the upright position, there is a danger that air may be sucked into the vein, and thence, from the right side of the heart to the lungs, producing an *air embolism*.
- **The fascial planes of the neck**: these are important to the surgeon as they provide useful planes of cleavage during dissection and govern the spread of pus in the neck. For example, tuberculous disease in the bodies of the cervical vertebrae may spread behind the prevertebral fascia to produce a *retropharyngeal abscess*, which may be visible as a bulge in the posterior wall of the pharynx.

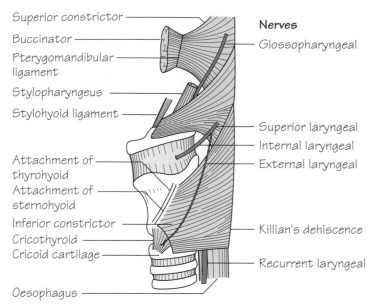

Superior constrictor
Buccinator
Pterygomandibular ligament
Stylopharyngeus
Stylohyoid ligament
Attachment of thyrohyoid
Attachment of sternohyoid
Inferior constrictor
Cricothyroid
Cricoid cartilage
Oesophagus

Nerves
Glossopharyngeal
Superior laryngeal
Internal laryngeal
External laryngeal
Killian's dehiscence
Recurrent laryngeal

Fig.67.1
The pharynx and larynx, and some of the related nerves

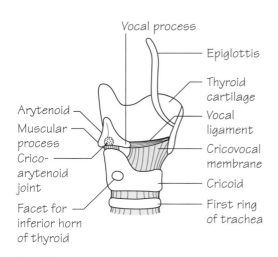

Vocal process
Epiglottis
Thyroid cartilage
Arytenoid
Muscular process
Crico-arytenoid joint
Facet for inferior horn of thyroid
Vocal ligament
Cricovocal membrane
Cricoid
First ring of trachea

Fig.67.2
A midline section of the larynx to show the cricovocal membrane and vocal ligaments

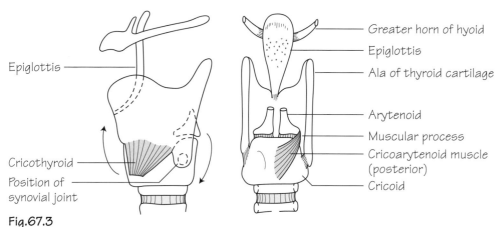

Epiglottis
Cricothyroid
Position of synovial joint

Greater horn of hyoid
Epiglottis
Ala of thyroid cartilage
Arytenoid
Muscular process
Cricoarytenoid muscle (posterior)
Cricoid

Fig.67.3
Left, the cricothyroid muscle and right, the posterior cricoarytenoid

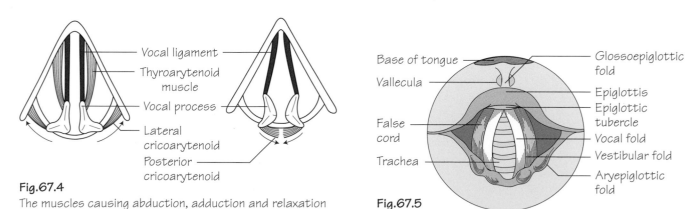

Vocal ligament
Thyroarytenoid muscle
Vocal process
Lateral cricoarytenoid
Posterior cricoarytenoid

Fig.67.4
The muscles causing abduction, adduction and relaxation of the vocal cords. Tightening of the cords is shown in Fig.67.3

Base of tongue
Vallecula
False cord
Trachea
Glossoepiglottic fold
Epiglottis
Epiglottic tubercle
Vocal fold
Vestibular fold
Aryepiglottic fold

Fig.67.5
The larynx as seen through a laryngoscope

 Anatomy at a Glance, Third Edition. Omar Faiz, Simon Blackburn and David Moffat.

The pharynx

This is an incomplete striated muscular tube, opening in front into the nasal cavity, the mouth and the larynx, thus being made up of the *nasopharynx*, the *oropharynx* and the *laryngopharynx*. The muscular coat (Fig. 67.1) is formed by:

- **The superior constrictor**: arises from the *pterygomandibular ligament* (which spans between the pterygoid hamulus and the mandible just behind the third molar tooth).
- **The middle constrictor**: arises from the *stylohyoid ligament* and the lesser and greater horns of the hyoid bone.
- **The inferior constrictor**: arises from the *thyroid* and *cricoid cartilages*. The fibres of the lower part of the muscle which arise from the cricoid are horizontal (sometimes called the *cricopharyngeus*) but, above this level, the fibres become more and more vertical. There is a triangular small gap between the two parts of the muscle (*Killian's dehiscence*, Fig. 67.1).

The constrictors encircle the pharynx and interdigitate posteriorly. The gaps between the constrictors are filled in by fascia. There is also an inner longitudinal layer of muscle. The *nasopharynx* is lined by ciliated columnar epithelium and, on its posterior wall, is situated a mass of lymphatic tissue, the *pharyngeal tonsil* or *adenoid*. The *pharyngotympanic (Eustachian) tube* opens into the nasopharynx at the level of the floor of the nose, the cartilage of the tube producing a distinct bulge behind the opening.

- **Nerve supply**:
  - *Motor supply*: pharyngeal branch of the vagus (p. 145).
  - *Sensory supply*: glossopharyngeal (p. 145).

The larynx
Palpable components
- **Hyoid bone**: level of C3.
- **Thyroid cartilage**: level of C4 and C5.
- **Cricoid cartilage**: level of C6.

Other components
- **Arytenoid cartilages**: attached to the upper border of the cricoid by synovial joints so that they can slide and rotate. Each has an anterior *vocal process* and a lateral *muscular process* (Figs. 67.2 and 67.3).
- **Epiglottis**: a leaf-shaped piece of elastic cartilage attached to the back of the thyroid cartilage (Fig. 67.2) and projecting upwards behind the hyoid.
- **Thyrohyoid ligament**: joins the hyoid and the thyroid (Fig. 67.1).
- **Cricovocal membrane (cricothyroid ligament)**: attached to the upper border of the cricoid, and passes inside the thyroid to be attached to the back of the thyroid and to the vocal processes of the arytenoids (Fig. 67.2). The upper border is thickened to form the *vocal ligament* which, with the mucous membrane that covers it, forms the *vocal cords*.
- **Cricothyroid joint**: a small synovial joint between the inferior horn of the thyroid cartilage and the cricoid, and permits a hinge-like movement (Fig. 67.3).
- **Pyriform fossa**: the fossa between the posterior border of the thyroid cartilage and the cricoid and arytenoid cartilages.
- **Mucous membrane**: mostly respiratory epithelium (ciliated columnar), but over the vocal cords it changes to stratified squamous so that the cords have a pearly appearance (Fig. 67.5).

- **Vestibular folds**: a pair of additional folds above the vocal folds (*false cords*). The space between the vocal and vestibular folds is the *sinus of the larynx*.

The intrinsic muscles of the larynx
- **Cricothyroid (Figs. 67.1 and 67.3)**: situated on the *outside* of the larynx and tenses the vocal cords.
- **Thyroarytenoid (Fig. 67.4)**: from the back of the thyroid cartilage to the vocal process of the arytenoid and relaxes the vocal cords.
- **Posterior cricoarytenoid (Figs. 67.3 and 67.4)**: from the back of the cricoid to the muscular process of the arytenoid. The posterior cricoarytenoid abducts the vocal cords and is the only muscle of the larynx to do so.
- **Lateral cricoarytenoid (Fig. 67.4)**: adducts the vocal cords.
- **Interarytenoids and aryepiglottic muscle**: form a 'sphincter' together with the epiglottis in order to close off the entrance to the larynx (*glottis*) during swallowing.
- **Nerve supply**:
  - *Motor supply*: recurrent laryngeal nerve, except for the cricothyroid, which is supplied by the *external laryngeal nerve*.
  - *Sensory supply*: above the vocal cords, the *internal laryngeal nerve* which enters the larynx through the thyrohyoid membrane. Below the vocal cords, the *recurrent laryngeal nerve*, (which is, therefore, a mixed nerve) which enters the larynx just behind the cricothyroid joint.

Clinical notes

- **Pharyngeal diverticulum**: Killian's dehiscence is a potentially weak spot in the pharyngeal wall, and it is possible for a pouch of mucous membrane to herniate through this gap. Such a diverticulum will fill with swallowed food and may cause obstruction by pressing on the outside of the pharynx. Its contents may spill into the larynx and be inhaled into the trachea and bronchi.
- **Laryngoscopy (Fig. 67.5)**: the vocal cords may be seen by means of a warmed, angled mirror passed to the back of the throat or by direct laryngoscopy. Only the edges of the cords can be seen because of the shadow of the false cords. Adduction of the cords may be observed by asking the patient to say a high-pitched 'eeeee', while abduction occurs when saying 'ahhh'. If one of the recurrent laryngeal nerves is divided, the cord lies in a position midway between adduction and abduction, but this does not produce very severe voice changes because the uninjured cord can cross the midline to reach the paralysed cord. If both nerves are cut, however, the cords lie in the *cadaveric position*, i.e. in the mid-position. If both nerves are damaged but not completely divided, the cords are adducted as the abductors are more affected than the adductors (*Semon's law*). This may cause breathing difficulties.
- **Tumours of the larynx**: the pyriform fossa is often known as the *silent area* of the larynx because it is not closely related to any important structures; a carcinoma in this region can remain symptomless until it has spread to the cervical lymph nodes, whereas a tumour of the vocal cords, however small, produces early voice changes.

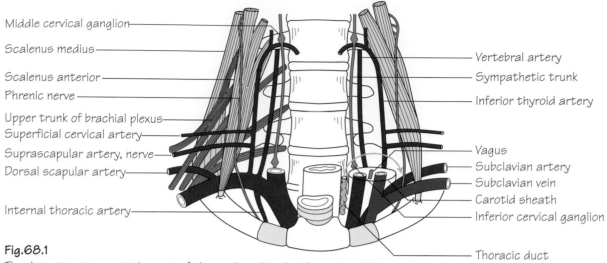

Middle cervical ganglion

Scalenus medius

Scalenus anterior

Phrenic nerve

Upper trunk of brachial plexus

Superficial cervical artery

Suprascapular artery, nerve

Dorsal scapular artery

Internal thoracic artery

Vertebral artery

Sympathetic trunk

Inferior thyroid artery

Vagus

Subclavian artery

Subclavian vein

Carotid sheath

Inferior cervical ganglion

Thoracic duct

Fig.68.1

The deep structures at the root of the neck and in the thoracic outlet.
The curved arrow on the right side of the diagram indicates the course of the thoracic duct

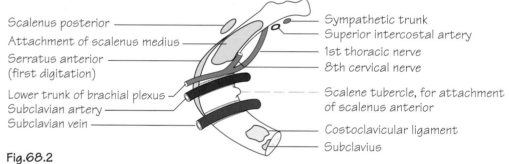

Scalenus posterior

Attachment of scalenus medius

Serratus anterior
(first digitation)

Lower trunk of brachial plexus

Subclavian artery

Subclavian vein

Sympathetic trunk

Superior intercostal artery

1st thoracic nerve

8th cervical nerve

Scalene tubercle, for attachment
of scalenus anterior

Costoclavicular ligament

Subclavius

Fig.68.2

The structures related to the
upper surface of the first rib

The area defined by the 1st thoracic vertebra, the 1st ribs and the manubrium sterni is called the *thoracic inlet (or outlet)* (Fig. 68.1). Through this relatively confined space pass the trachea and oesophagus, the carotid and subclavian arteries and the corresponding large veins as well as the apices of the lungs and important nerves.

The scalene muscles

- **Scalenus anterior**: passes down from the transverse processes of some of the cervical vertebrae and is inserted by means of a narrow tendon into the scalene tubercle on the medial border of the 1st rib.
- **Scalenus medius**: is behind the scalenus anterior and is inserted by muscular fibres into a large area of the 1st rib. The subclavian artery and the trunks of the brachial plexus are between the two muscles and the subclavian vein is anterior to scalenus anterior.

The arteries

- **The subclavian artery** (p. 149): on the right side, it arises from the *brachiocephalic artery* and, on the left, directly from the *arch of the aorta*. It arches over the apex of the lung and crosses the 1st rib in a shallow groove, which it shares with the lower trunk of the brachial plexus (Fig. 68.2). At the outer border of the 1st rib it becomes the axillary artery. It has five branches (p. 149):
  - The *vertebral artery*.
  - The *internal thoracic artery*.
  - The *thyrocervical trunk*.
  - The *costocervical trunk*.
  - The *dorsal scapular artery*.

The veins

- **The subclavian vein**: begins at the outer border of the 1st rib and lies in a shallow groove on the superior surface of the rib anterior to scalenus anterior. At the medial border of this muscle, it is joined by the *internal jugular vein* to form the *brachiocephalic vein*. The internal jugular vein is enclosed in the carotid sheath, along with the common carotid artery and the vagus nerve. Other veins entering it accompany the small arteries, but the inferior thyroid veins are solitary and run down from the lower border of the thyroid gland, in front of the trachea, to reach the left brachiocephalic vein in the thorax.

The nerves

- **The upper, middle and lower trunks of the brachial plexus**: emerge from between scalenus anterior and medius and pass down into the axilla. The suprascapular nerve arises from the upper trunk and joins the corresponding artery before passing back to the suprascapular notch. For further details, see Chapter 35.
- **The phrenic nerve**: is formed by branches from the 3rd, 4th (predominantly) and 5th cervical nerves and descends on the anterior surface of scalenus anterior before crossing the subclavian artery and entering the thorax.
- **The vagus nerve**: this crosses the subclavian artery and descends into the thorax. On the right side it gives off the right recurrent laryngeal nerve, which hooks under the artery and ascends, deep to the common carotid, to reach the larynx. The left recurrent laryngeal nerve, having arisen in the thorax, runs upwards between the trachea and oesophagus.
- **The sympathetic trunk**: descends close to the vertebral artery. The *middle cervical ganglion* is close to the entry of the artery into the foramen transversarium of C6 and the *inferior cervical ganglion* is near the neck of the 1st rib behind the origin of the vertebral artery. It may be fused with the 1st thoracic ganglion to form the *stellate ganglion*.

The thoracic duct (Fig. 68.1)

On the left side only, the duct ascends out of the thorax between the trachea and oesophagus, and arches laterally between the carotid sheath in front and the vertebral artery behind. It ends by joining the junction between the internal jugular and subclavian veins. On the right side, the jugular, subclavian and mediastinal lymph trunks usually unite to form the *right lymph duct* that opens into the right subclavian vein, but they may open independently.

Clinical notes

- **Thoracic outlet syndrome**: the triangular gap between scalenus anterior and medius may be narrow, and the tough tendon of scalenus anterior may compress the lower trunk of the brachial plexus giving rise to wasting of the small muscles of the hand and loss of, or diminished, sensation in the corresponding dermatomes. The subclavian artery may be compressed causing a cold blue hand. There may also be a *post-stenotic dilatation* of the artery in which thrombi may occur. This combination of symptoms is sometimes referred to as the *thoracic outlet syndrome* and, it can also be caused by a cervical rib (Chapter 3), tumours at the apex of the lung or enlarged lymph nodes.

69 The oesophagus and trachea and the thyroid gland

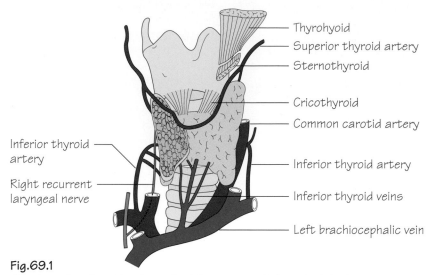

Fig.69.1
The thyroid and its blood supply. A large part of the right lobe has been removed

Labels:
- Thyrohyoid
- Superior thyroid artery
- Sternothyroid
- Cricothyroid
- Common carotid artery
- Inferior thyroid artery
- Inferior thyroid veins
- Left brachiocephalic vein
- Inferior thyroid artery
- Right recurrent laryngeal nerve

The oesophagus

The oesophagus begins at the level of the cricoid cartilage and runs down behind and slightly to the left of the trachea. The *left recurrent laryngeal nerve* is in the groove between the oesophagus and trachea and the *thoracic duct* is to the left of the oesophagus. The oesophagus is lined by stratified squamous ('wear and tear') epithelium.

The trachea

The trachea begins at the level of the cricoid cartilage (C6) and ends by dividing into left and right main bronchi at the level of the manubriosternal joint (lower border of T4). It is held open by its cartilaginous rings. As the larynx ascends on swallowing and the diaphragm descends during inspiration, the trachea is necessarily extensible. It can be palpated in the midline just above the suprasternal notch and can be seen on an X-ray as a dark shadow. The 2nd, 3rd and 4th rings of the trachea are crossed by the isthmus of the thyroid gland.

The infrahyoid ('strap') muscles

• **Sternothyroid**: arises from the back of the manubrium and ascends to be attached to the outer surface of the thyroid cartilage.
• **Thyrohyoid**: is a continuation of the latter and is attached to the hyoid bone.
• **Sternohyoid**: is superficial to the other two and runs from the manubrium to the lower border of the hyoid.
• **Omohyoid**: the superior belly is attached to the hyoid and runs down to its intermediate tendon, and then continues as the inferior belly across the posterior triangle to be attached to the scapula.

These infrahyoid muscles are all supplied by the *ansa cervicalis* (C1, 2 and 3). Their function is to fix the hyoid bone so that the suprahyoid muscles have a fixed base. Their main importance lies in their close relation to the thyroid gland.

The thyroid gland

The thyroid is an endocrine gland with an extremely rich blood supply (Fig. 69.1). Its isthmus lies across the 2nd, 3rd and 4th rings of the trachea and the lobes lie on either side, reaching up as far as the 'pocket' under the attachment of the sternothyroid to the thyroid cartilage. There may be a small *pyramidal lobe* above the isthmus, attached to the body of the hyoid bone by a fibromuscular band. The thyroid is enclosed in the thin *pretracheal fascia*, and it has its own fibrous capsule.

• **Blood supply**:
  • The *superior thyroid artery*: comes from the external carotid and runs down to enter the 'pocket' between the sternothyroid and the thyroid cartilage where it is close to the external laryngeal nerve. It divides into two branches which run down the posterior border and along the upper border.
  • The *inferior thyroid artery*: has been described above (p. 149), and there is a very free anastomosis between the two arteries, their branches and the small arteries of the trachea and oesophagus. It is, thus, possible to tie all four arteries during subtotal thyroidectomy and still leave an adequate blood supply to the remaining thyroid tissue and the parathyroids.
• **Venous drainage**: there are three veins on each side: the *superior* and *middle thyroid veins* drain into the internal jugular, and the *inferior thyroid veins* drain downwards into the mediastinum to open into the left brachiocephalic vein.

The parathyroid glands

There are two parathyroids, superior and inferior, on each side. They are about the size of a pea and are embedded in the back of the thyroid (but outside its capsule). The superior parathyroid is fairly constant in position, but the inferior gland is rather variable and may lie lower down than its usual position near the lower pole of the thyroid, and may even be found in the mediastinum near the thymus gland (see Chapter 76).

Clinical notes

- **Oesophageal constrictions**: there are four slight constrictions in the oesophagus where swallowed foreign bodies are liable to be held up or where there may be obstruction to the passage of instruments. These are at the level of the cricopharyngeus muscle, where it is crossed by the arch of the aorta, where it is crossed by the left bronchus and where it pierces the diaphragm. These constrictions are 15, 22, 27 and 40 cm, respectively, from the incisor teeth.
- **Oesophagitis**: the stratified squamous epithelium of the oesophagus is not resistant to acidic gastric juice in the same way as the columnar mucous-secreting epithelium of the stomach. If there is gastro-oesophageal reflux, such as may occur in a sliding hiatus hernia, there may be inflammation and ulceration of the lower oesophagus.
- **Tracheostomy**: this is indicated when there is some obstruction to the larynx (for example by inhaled foreign bodies) or for long-lasting artificial ventilation. The trachea may be exposed in the midline below the cricoid cartilage. The pretracheal fascia is divided and the isthmus of the thyroid is pushed upwards or divided between clamps, avoiding the inferior thyroid veins. The trachea is opened and a tracheostomy tube is inserted. The operation is difficult in children owing to the very small size of the trachea (about 3 mm in diameter during the first year), the softness of the tracheal rings and the fact that the left brachiocephalic vein may be above the suprasternal notch in young children.
- **Goitre**: this is the term used for any enlargement of the thyroid gland. When the thyroid becomes enlarged, the strap muscles are stretched tightly over it, and it may displace the carotid sheath and its contents laterally. It may compress the oesophagus and trachea. As the thyroid is enclosed in the pretracheal fascia, and this is attached to the larynx above, swellings of the thyroid move upwards on swallowing.

  In the operation of *subtotal thyroidectomy*, the four main arteries are divided, avoiding the associated nerves (*external laryngeal* for the superior thyroid artery and *recurrent laryngeal* for the inferior), and the gland is then removed except for a narrow strip posteriorly lying alongside the oesophagus and trachea. The parathyroids are thus left *in situ*. The blood supply of this remnant is by means of the anastomoses between the thyroid vessels and the vessels of the oesophagus and trachea.
- **Thyroglossal remnants**: cysts or aberrant thyroid tissue may be found along the course of the thyroglossal duct anywhere between the thyroid and the foramen caecum of the tongue (Chapter 76) and may need removal, together with remnants of the duct itself and the central part of the body of the hyoid bone.
- **Parathyroid tumours**: parathyroid tumours may need surgical removal because of their production of excessive amounts of parathyroid hormone (*hyperparathyroidism*). They are sought at the posterior border of the thyroid gland but, as mentioned above, the inferior gland may be difficult to find.

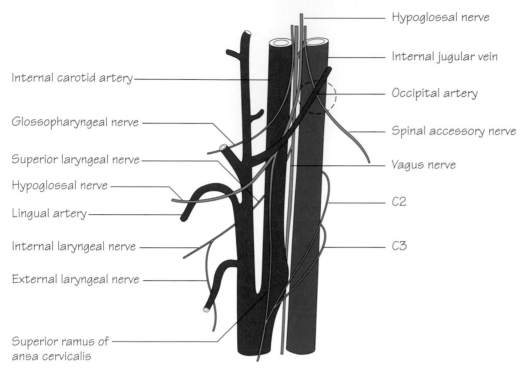

Internal carotid artery

Glossopharyngeal nerve

Superior laryngeal nerve

Hypoglossal nerve

Lingual artery

Internal laryngeal nerve

External laryngeal nerve

Superior ramus of ansa cervicalis

Hypoglossal nerve

Internal jugular vein

Occipital artery

Spinal accessory nerve

Vagus nerve

C2

C3

Fig.70.1
The last four cranial nerves and their relation to the large vessels

Disc

Head of mandible

Lateral pterygoid

Styloid process

Fibrocartilage

External auditory meatus

Mastoid process

Fig.70.2
The temporomandibular joint

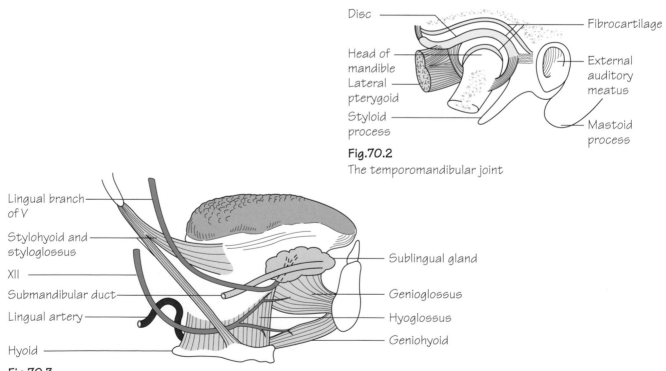

Lingual branch of V

Stylohyoid and styloglossus

XII

Submandibular duct

Lingual artery

Hyoid

Sublingual gland

Genioglossus

Hyoglossus

Geniohyoid

Fig.70.3
The extrinsic muscle of the tongue and the related nerves

The 'foundation' for the upper part of the neck consists of the *superior* and *middle constrictors* of the pharynx, on which lie the *internal* and *external carotid arteries*, the *internal jugular vein* and the *last four cranial nerves* (Fig. 70.1).
- **The glossopharyngeal**: runs forwards and across the internal carotid artery, but deep to the external carotid (p. 145).
- **The vagus**: which is joined by the *cranial root of the accessory* and runs straight down, between the internal carotid and the internal jugular, and within the carotid sheath (p. 145).
- **The spinal root of the accessory**: runs backwards, crossing the internal jugular vein and the transverse process of the atlas, to supply sternomastoid and trapezius (p. 145).
- **The hypoglossal**: having left the cranial cavity by means of the hypoglossal canal, it is joined by the anterior ramus of the 1st cervical nerve and winds round the vagus, and then runs downwards and forwards, superficial to both carotids, giving off the *descendens hypoglossi* in the process. The branches and distribution of these nerves have already been described (p. 145).

The infratemporal region

This is the region deep to the ramus of the mandible.

Contents
- **The stem of the mandibular division of the trigeminal nerve**: which enters through the foramen ovale and immediately breaks up into branches (Chapter 62).
- **The otic ganglion**: which lies medial to the nerve (Chapter 62).
- **the medial and lateral pterygoid muscles**: the *medial pterygoid* is inserted into the inner surface of the ramus and thus separates the region from the structures lying on the superior constrictor described above. The *lateral pterygoid* runs backwards from the lateral pterygoid plate to the neck of the mandible and the intra-articular disc.
- **The maxillary artery**: enters the region by passing forwards deep to the neck of the mandible, and its branches correspond to those of the nerve, with the addition of the **middle meningeal artery** which ascends to pass through the *foramen spinosum*. The maxillary artery leaves by entering the *pterygopalatine fossa*.
- **The temporomandibular joint** (Fig. 70.2): this is a synovial joint with an intra-articular disc but, unlike most other synovial joints, the articular cartilage and the disc are composed of fibrocartilage or even fibrous tissue. The lateral pterygoid muscle can pull the disc and the head of the mandible forwards onto the articular eminence. This occurs when the mouth is opened, so that the joint is not a simple hinge joint. The axis of rotation is through the mandibular foramen, so that the inferior alveolar nerve and vessels are not stretched when the mouth is opened. The mouth is opened by gravity and the suprahyoid muscles such as *mylohyoid* and *geniohyoid* and closed by the *masseter*, *temporalis* and *medial pterygoid*.

The submandibular region

This is the region below the mandible and extending upwards deep to the mandible as far as the attachment of the *mylohyoid muscle* to the mylohyoid line of the mandible. The contents of the submandibular region include:

Muscles
- **Mylohyoid**: is attached to the hyoid and the mylohyoid line on the mandible (Fig. 60.2). On its surface lies the **anterior belly of the digastric muscle**, and the two have the same nerve supply (the *mylohyoid nerve*).
- **The posterior belly of the digastric**: runs back to a groove medial to the mastoid process, and the intermediate tendon is attached to the hyoid bone.
- **Hyoglossus**: runs from the greater horn of the hyoid up to the side of the tongue and is partly deep to the mylohyoid.
- **The middle constrictor of the pharynx**: lies behind and partly deep to the hyoglossus.

Nerves and vessels (Fig. 70.3)
- **The lingual artery**: leaves the external carotid with an upward loop and then runs forwards deep to the hyoglossus. The lingual artery supplies the tongue.
- **The lingual nerve**: enters the region by passing just behind the 3rd molar tooth, directly in contact with the mandible, and then loops forwards on the hyoglossus to enter the tongue. Suspended from the lingual nerve is the *submandibular ganglion*, in which parasympathetic fibres from the *chorda tympani* synapse before supplying the submandibular and sublingual glands. The lingual nerve carries *sensory fibres* from the anterior two-thirds of the tongue as well as *taste fibres* which are carried in the chorda tympani.
- **The hypoglossal nerve**: crosses the loop of the lingual artery and then runs forwards on the hyoglossus, below the lingual nerve, to enter the tongue and supply its intrinsic and extrinsic muscles.

Salivary glands
- **The submandibular gland**: lies on the mylohyoid and the anterior belly of the digastric, extending up as far as the mylohyoid line. It also extends back onto the hyoglossus and has a deep process which passes forwards, deep to the mylohyoid. From this the *submandibular (Wharton's) duct* travels forwards to enter the mouth at the *sublingual papilla* near the midline. The duct is crossed by the lingual nerve, which then passes deep to the duct to enter the tongue. The *facial artery* is embedded in the posterior part of the gland before turning down, between the gland and the inner surface of the mandible, and then passing over the lower border of the mandible to supply the face.
- **The sublingual gland**: lies deep to the mylohyoid, near to the midline. Its upper surface is covered by the mucous membrane of the mouth, and its numerous ducts open onto a ridge in the floor of the mouth extending back from the sublingual papilla.

Clinical notes
- **Dislocation of the mandible**: when the mouth is open, the head of the mandible is on the articular eminence. A blow on the jaw in this position may cause a dislocation of the mandible. It is reduced by using the (protected) thumbs to push down on the molar teeth in order to relieve the spasm of the temporalis and masseter muscles and then to push the mandible back into place.
- **Stones in the submandibular duct**: stones are liable to occur in the submandibular duct, causing obstruction, so that the gland swells up, especially when eating or even at the sight of food. The duct may be palpated by *bimanual examination* with a finger in the mouth and the fingers of the other hand outside. The stone may usually be removed.

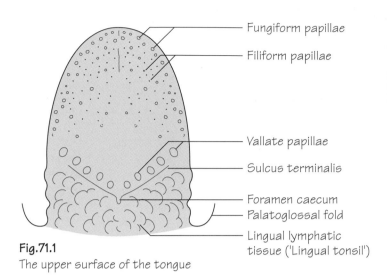

Fig.71.1

The upper surface of the tongue

- Fungiform papillae
- Filiform papillae
- Vallate papillae
- Sulcus terminalis
- Foramen caecum
- Palatoglossal fold
- Lingual lymphatic tissue ('Lingual tonsil')

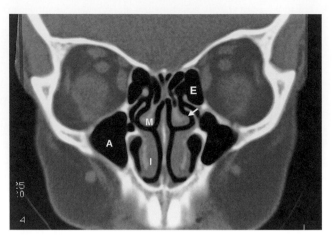

Fig.71.3

CT scan showing a coronal section through the paranasal sinuses. A, maxillary antrum; E, ethmoid sinus; L, inferior turbinate; M, middle turbinate. The arrow is pointing to the middle meatus

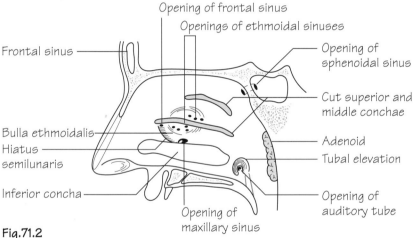

Fig.71.2

The lateral wall of the nasal cavity

- Opening of frontal sinus
- Openings of ethmoidal sinuses
- Frontal sinus
- Opening of sphenoidal sinus
- Cut superior and middle conchae
- Bulla ethmoidalis
- Hiatus semilunaris
- Adenoid
- Tubal elevation
- Inferior concha
- Opening of auditory tube
- Opening of maxillary sinus

The *hard palate* is formed by the *palatal process of the maxilla* and the *horizontal plate of the palatine bone*. The *soft palate* hangs like a curtain behind the mouth cavity.

Muscles
- **Levator palati**: elevates the palate.
- **Tensor palati**: tenses the palate. These two muscles move the soft palate towards the back wall of the oropharynx where it meets a part of the *superior constrictor* which contracts strongly to form a ridge – the *ridge of Passavant*. The mouth and nasal cavities are, thus, separated so that food does not regurgitate into the nose, and so as to be able to pronounce the palatal consonants, such as 'g' and 'k'.

- **Palatoglossus and palatopharyngeus**: these pass downwards from the palate to the side of the tongue and the inside of the pharynx, respectively. They raise two ridges, the *palatoglossal* and *palatopharyngeal arches*, that are also called the *anterior* and *posterior pillars of the fauces*. They separate the mouth from the oropharynx and between them is the *tonsillar fossa*.

The nerve supply of the pharynx and soft palate
The *pharyngeal plexus* is a plexus of nerves formed by:
- **The pharyngeal branch of the vagus**: which includes the cranial root of the accessory. It provides the motor supply to the muscles, except for the tensor palati which is supplied by the mandibular division of the trigeminal.

- **The glossopharyngeal nerve**: which provides the sensory supply to the pharynx.
- **Branches from the sympathetic trunk**.

The sensory nerve supply of the palate

The palate is supplied by the greater and lesser palatine and the nasopalatine nerves from the maxillary division of the trigeminal (see Fig. 62.2). These nerves also supply the inner surface of the gums.

Other features

- **The tonsil**: a mass of lymphatic tissue lying in the tonsillar fossa which, like the rest of the lymphatic system, reaches its maximum size at puberty. Lateral to the tonsil is its fibrous capsule and the superior constrictor. It is supplied by the *tonsillar branch of the facial artery*. The bleeding that occurs after tonsillectomy, however, is usually from the *paratonsillar vein*. The *pharyngeal tonsil* (*adenoid*) has already been mentioned, and there is also a *lingual tonsil* lying in the back of the tongue.
- **The teeth**: the *deciduous (milk) teeth* comprise two *incisors*, a *canine* and two *molars*; the last occupy the position of the two premolars of the permanent teeth. The *permanent teeth* comprise two *incisors*, a *canine*, two *premolars* and three *molars*. The first milk teeth to erupt are usually the lower central incisors at about *6 months* and the first permanent teeth are the first molars at about *6 years*.

The tongue

The tongue is divided, developmentally and anatomically, into an anterior two-thirds and a posterior one-third, separated by the *sulcus terminalis*, a V-shaped groove with the foramen caecum at the apex (Fig. 71.1). The latter is the site of outgrowth of the *thyroglossal duct* (see Chapter 76). In front of the sulcus is a row of *vallate papillae*. *Filiform papillae* and red, flat-topped *fungiform papillae* can be seen on the more anterior parts of the tongue.

Muscles (Fig. 70.3)

- **Intrinsic muscles**: run in three directions, longitudinal, transverse and vertical.
- **Hyoglossus**: from the greater horn of the hyoid bone.
- **Genioglossus**: from the genial tubercle on the back of the mandible.
- **Styloglossus**: from the styloid process.

The last three muscles blend in with the intrinsic muscles. Genioglossus is especially important as it is inserted along the whole length of the tongue so that it is used to protrude the tongue.

Nerve supply

- **Motor supply**: hypoglossal nerve.
- **Sensory supply**: anterior two-thirds by the *lingual nerve*; taste fibres travel in the *chorda tympani*. Posterior one-third by the *glossopharyngeal nerve*. A small part of the tongue near the epiglottis is supplied by the *internal laryngeal branch* of the vagus nerve. The nerve supply is the result of the development of the tongue from the branchial arches (Chapter 76).

The nasal cavity

The boundaries of the nasal cavity include the:
- **Nasal septum**: the perpendicular plate of the ethmoid, vomer and a large plate of cartilage.
- **Lateral wall (Fig. 71.2)**: maxilla, lacrimal, ethmoid with its superior and middle conchae, perpendicular plate of the palatine and inferior concha (a separate bone).
- **Roof**: nasal bones, cribriform plate of the ethmoid and body of the sphenoid.
- **Floor**: palatal processes of the maxilla and palatine bones.

The spaces beneath the conchae are the *meatuses* and the region above the superior concha is the *spheno-ethmoidal recess*.

The paranasal sinuses (Fig. 71.3)

- **The maxillary sinus**: is situated inside the body of the maxilla and opens into the middle meatus. Most of the sinus is below the level of the opening.
- **The frontal sinuses**: on each side of the midline, just above the medial part of the orbit. The frontal sinuses drain into the middle meatus.
- **The ethmoidal sinuses**: in the body of the ethmoid bone and are, therefore, deep to the medial wall of the orbit. They drain into the middle and superior meatuses.
- **The sphenoidal sinus**: inside the body of the sphenoid and drains into the spheno-ethmoidal recess.

The nasolacrimal duct

The nasolacrimal duct drains tears from the medial angle of the eye via the lacrimal puncta. It drains into the inferior meatus.

Clinical notes

- **Tongue incisions**: because the tongue largely develops from a pair of lingual swellings (Chapter 76), the nerves and blood vessels do not cross the midline, although a few lymphatics do, so that a midline incision cannot damage any important structures.
- **Lesions of the hypoglossal nerve**: a lesion of the hypoglossal nucleus in the medulla, or of the nerve itself, will produce unilateral wasting of the tongue musculature. When the patient is asked to protrude his or her tongue, it will deviate to the side of the lesion owing to the paralysis of genioglossus.
- **Sinusitis**: upper respiratory tract infection can give rise to *sinusitis*; the maxillary sinus can be infected from the roots of the teeth which are closely related to the floor of the sinus. This sinus is particularly prone to chronic infection because its opening into the middle meatus is at the top of the sinus, so that gravitational drainage cannot occur.
- **Nasolacrimal obstruction**: the nasolacrimal duct may become blocked, leading to overflow of tears from the eye. It can be cannulated through one of the lacrimal puncta (Chapter 72) and washed out with saline.

The face and scalp

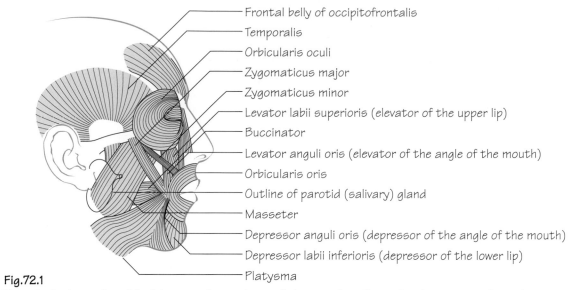

Frontal belly of occipitofrontalis
Temporalis
Orbicularis oculi
Zygomaticus major
Zygomaticus minor
Levator labii superioris (elevator of the upper lip)
Buccinator
Levator anguli oris (elevator of the angle of the mouth)
Orbicularis oris
Outline of parotid (salivary) gland
Masseter
Depressor anguli oris (depressor of the angle of the mouth)
Depressor labii inferioris (depressor of the lower lip)
Platysma

Fig.72.1
The principal muscles of facial expression and two of the muscles of mastication, temporalis and masseter

The facial muscles

• **Muscles of mastication**: see Muscle index, p. 183. They are all supplied by the mandibular division of the trigeminal (p. 143).

• **Muscles of facial expression**: they are all supplied by the facial nerve. They have only one attachment to bone, or sometimes no attachment at all, the other end of the muscle being inserted into skin or blending with other muscles. The most important are shown in Fig. 72.1, from which their actions can be deduced. The *orbicularis oculi* has an *orbital* part which surrounds the eye as a sphincter and closes it tightly, while the *palpebral* part, which is in the eyelid, closes the eye gently as in sleep. The most important function of the *buccinator* is to keep the cheeks in contact with the gums so that food does not collect in this region. The *platysma* extends down the neck and over the clavicle and upper part of the chest.

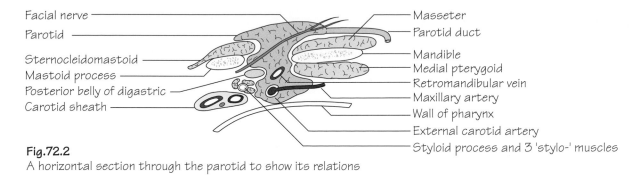

Fig.72.2
A horizontal section through the parotid to show its relations

Labels (left, top to bottom): Facial nerve, Parotid, Sternocleidomastoid, Mastoid process, Posterior belly of digastric, Carotid sheath

Labels (right, top to bottom): Masseter, Parotid duct, Mandible, Medial pterygoid, Retromandibular vein, Maxillary artery, Wall of pharynx, External carotid artery, Styloid process and 3 'stylo-' muscles

The parotid gland

Situated mainly behind the mandible but spills over it onto the face. It extends deeply to come into contact with the pharynx, and is moulded around the mastoid process and sternomastoid posteriorly. The *parotid duct* extends forwards across masseter to enter the mouth opposite the second upper molar. The whole gland is enclosed in dense fascia so that swelling of the gland, as in mumps for instance, is very painful. It is separated from the submandibular gland by a thickening in the fibrous capsule called the *stylomandibular ligament*. Three structures pass through the gland (Fig. 72.2). These are, from superficial to deep, the *facial nerve*, the *retromandibular vein* (the beginning of the external jugular) and the *external carotid artery*, with its *maxillary* and *superficial temporal* branches.

Nerves of the face

• **The facial nerve**: having left the stylomastoid foramen, it enters the parotid and divides into *temporal*, *zygomatic*, *buccal*, *marginal mandibular* and *cervical* branches (Fig. 72.3), with some intercommunicating branches between them. Note that the marginal mandibular branch lies below the mandible for part of its course (Fig. 63.1), so that submandibular incisions are made well below the mandible to avoid cutting this nerve. The cervical branch supplies platysma.

• **The trigeminal nerve**: sensory to the whole face (Fig. 72.3) except for the area over the parotid (Chapter 62).

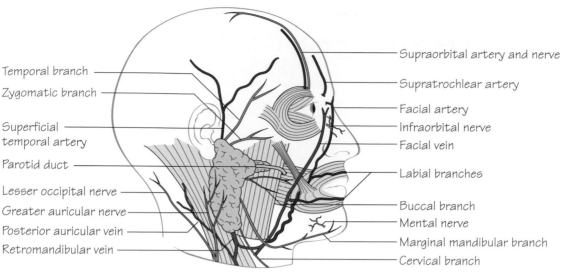

Fig.72.3
The principal nerves and blood vessels of the face

Labels:
Temporal branch
Zygomatic branch
Superficial temporal artery
Parotid duct
Lesser occipital nerve
Greater auricular nerve
Posterior auricular vein
Retromandibular vein

Supraorbital artery and nerve
Supratrochlear artery
Facial artery
Infraorbital nerve
Facial vein
Labial branches
Buccal branch
Mental nerve
Marginal mandibular branch
Cervical branch

Blood vessels of the face (Fig. 72.3)

• **The facial artery (see p. 147)**: enters the face by passing over the lower border of the mandible at the anterior border of masseter. It has a tortuous course, passing close to the corner of the mouth and then alongside the nose to end near the medial angle of the eye. The facial artery anastomoses freely across the midline and with other arteries on the face.

• **The facial vein**: follows a straighter path than the artery and anastomoses, at the medial angle of the eye, with the ophthalmic veins and, thus, indirectly, with the cavernous sinus.

The eye

• **The conjunctiva**: covers the surface of the eye and is reflected onto the inner surface of the eyelids, the angle of reflection forming the *fornix* of the conjunctival sac.

• **The tarsal plates**: are composed of dense fibrous tissue, more compact in the upper than the lower eyelid. Outside these are the muscle fibres of the palpebral part of the orbicularis oculi, some loose areolar tissue and skin. Partly embedded in the deep surface of the tarsal plates are the *tarsal (Meibomian) glands*, which open onto the edge of the eyelids and produce a modified form of sebum.

• **The lacrimal gland**: is in the upper lateral part of the orbit, lying in a shallow hollow in the bone. Its secretions pass through 9–12 ducts into the superior fornix of the conjunctiva and thence across the eye to the medial angle (*canthus*). From here, the tears pass into the *lacrimal puncta*, two minute openings in the upper and lower eyelids, and thence into the *lacrimal sac* lying in a groove in the lacrimal bone. This drains the tears into the *nasolacrimal duct*, which opens into the inferior meatus of the nose.

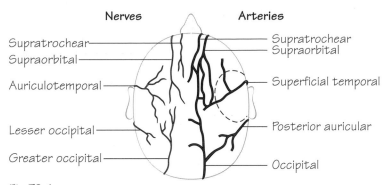

Nerves   **Arteries**

Supratrochear ——————   —————— Supratrochear
Supraorbital ——————   —————— Supraorbital

Auriculotemporal ——————   —————— Superficial temporal

Lesser occipital ——————   —————— Posterior auricular

Greater occipital ——————   —————— Occipital

Fig.72.4
The nerves and blood vessels of the scalp.
The dotted line shows a temporal 'flap'

The scalp

The scalp is made up of five layers, which form a useful mnemonic:
- **S**kin.
- **C**utaneous fat and connective tissue.
- **A**poneurosis (epicranial): this is a tough sheet of dense connective tissue into which are inserted the occipital and frontal bellies of the *occipitofrontalis* muscle.
- **L**oose areolar tissue: this is a mobile layer which allows the main part of the scalp to move freely over the surface of the skull.
- **P**ericranium: the periosteum of the outside of the skull.

Blood vessels and nerves

All blood vessels and nerves enter from the periphery and are shown in Fig. 72.4. The vessels anastomose freely, and the arteries are embedded in dense connective tissue and cannot retract, so that bleeding is copious in injuries to the scalp but can always be stopped by direct pressure on the wound.

Emissary veins

These are small veins that pass through the skull and unite the veins of the scalp with the intracranial veins. They form a possible route for infection to reach the cranial cavity.

Clinical notes

- **Facial nerve lesions**: the facial nerve may be affected by tumours in the cerebello-pontine angle and fractures of the base of the skull, in the course of the nerve through the stylomastoid canal or on the face.
- **Bell's palsy**: see p. 145.
- **Parotid tumours**: the facial nerve may be involved in malignant tumours of the parotid.
- **Facial paralysis**: paralysis of the facial musculature as a result of any of the lesions mentioned leads to asymmetry of the face as the affected side droops because of the loss of muscle tone. Paralysis of the orbicularis oculi causes the lower eyelid to droop, so that tears may escape from the eye. The eye cannot be closed. Drooping of the corner of the mouth can result in dribbling of saliva and there is difficulty in speech (*dysarthria*). Paralysis of buccinator leads to food collecting in the space between the teeth and gums. On attempting to smile, the face is drawn over towards the sound side and whistling is impossible. The upper part of the face is less affected in *supranuclear (upper motor neurone) lesions* because it is bilaterally represented in the cerebral cortex.
- **Infections of the face**: infections of the central region of the face need early and effective treatment as there is a danger of the infection causing thrombosis of the facial vein and spreading to the cavernous sinus by means of the communication between the facial vein and the ophthalmic veins.
- **The scalp**: the layer of loose areolar tissue in the scalp is a potential plane of cleavage, so that a glancing blow to the head or long hair caught up in rapidly revolving machinery can cause partial or complete 'scalping'. Fortunately, the copious blood supply to the scalp and the free anastomoses between the vessels enable the avulsed scalp to be replaced, with an excellent prospect of complete recovery.

    Haemorrhage into the layer of areolar tissue can spread widely beneath the scalp down as far as the zygomatic arch and into the upper eyelids. Subconjunctival haemorrhages are bright red because the blood remains oxygenated.

    Surgical incisions for exposure of the brain are based on the arteries that enter the scalp from the periphery. For example, a *temporal flap* is based on the superficial temporal artery which keeps the flap viable until it is replaced. Similar flaps are based on the supraorbital and occipital arteries.

73 The cranial cavity

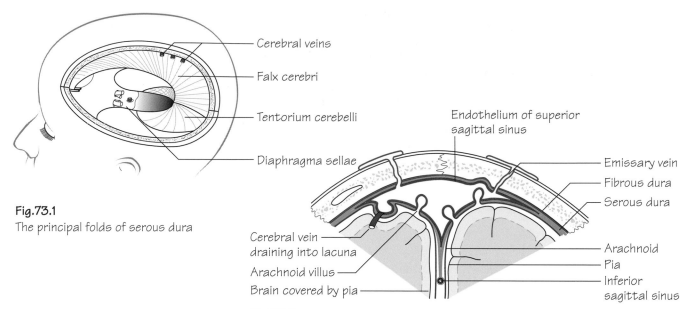

Cerebral veins

Falx cerebri

Tentorium cerebelli

Diaphragma sellae

Fig.73.1
The principal folds of serous dura

Endothelium of superior sagittal sinus

Emissary vein

Fibrous dura

Serous dura

Cerebral vein draining into lacuna

Arachnoid villus

Brain covered by pia

Arachnoid

Pia

Inferior sagittal sinus

Fig.73.2
A cross-section through the superior sagittal sinus to show the arachnoid villi projecting into the sinus

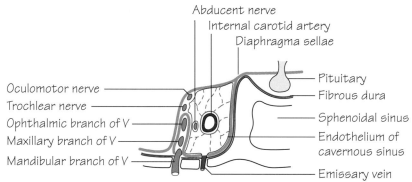

Abducent nerve

Internal carotid artery

Diaphragma sellae

Oculomotor nerve

Trochlear nerve

Ophthalmic branch of V

Maxillary branch of V

Mandibular branch of V

Pituitary

Fibrous dura

Sphenoidal sinus

Endothelium of cavernous sinus

Emissary vein

Fig.73.3
A cross-section through the cavernous sinus

The meninges

The meninges comprise the *dura*, *pia* and *arachnoid* mater. They surround the brain and, except for the fibrous layer of the dura, are continuous with the meninges of the spinal cord.

• **The dura mater (Fig. 73.1)**: comprises two layers which are fused, except where they separate to form the *dural venous sinuses*, which are lined by endothelium (Chapter 65). The *fibrous layer* is closely adherent to the bone and forms the periosteum of the cranial bones. The *serous layer* forms the dural venous sinuses and two large sheets – the *falx cerebri* and the *tentorium cerebelli* (see below).

• **The arachnoid mater**: lies deep to the dura and bridges over the sulci and fissures of the brain. The subarachnoid space contains the cerebrospinal fluid and the major blood vessels of the brain. It is continuous with the subarachnoid space of the spinal cord and ends at the level of the second sacral vertebra. It also surrounds the optic nerve as far as the back of the eyeball.

• **The pia mater**: is a thin delicate layer which follows the contours of the brain, dipping into the sulci.

The falx cerebri

The *falx cerebri* projects down in the sagittal plane and partially separates the two cerebral hemispheres. It tapers to a point anteriorly, but posteriorly is attached to the tentorium. The *superior sagittal sinus* (Fig. 73.2) is in its attached border and the *inferior sagittal sinus* is in its free border. Veins from the cerebral hemispheres drain into the superior sagittal sinus or into diverticula from it, the *lacunae laterales*. In places, the underlying arachnoid sends small outgrowths through the serous dura to project into the sinus. These are the *arachnoid villi* and are the site of absorption of cerebrospinal fluid into the bloodstream. In later life, they clump together to form *arachnoid granulations* which make indentations in the skull.

The tentorium cerebelli

The *tentorium cerebelli* forms a roof over the posterior cranial fossa and the cerebellum, transmitting the brainstem through an anterior, midline, hole. It is attached to the posterior clinoid processes, the petrous temporal bone and the inside of the skull. Its two layers split to enclose the transverse sinuses.

There are two smaller projections of serous dura – the *falx cerebelli* between the two cerebellar hemispheres and the *diaphragma sellae* which forms a roof over the pituitary fossa and the pituitary gland.

The cavernous sinus (Fig. 73.3)

The cavernous sinus lies on either side of the pituitary fossa and the body of the sphenoid. Like the other venous sinuses, it is formed by a layer of serous dura lined by endothelium. In addition, a layer of serous dura from the posterior cranial fossa projects forwards into the side of the cavernous sinus to form the *cavum trigeminale (Meckel's cave)* which contains the trigeminal ganglion and the proximal parts of its three branches.

Contents of the cavernous sinus

• **Internal carotid artery**: see p. 147.
• **Oculomotor nerve**: see p. 141.
• **Trochlear nerve**: see p. 141.
• **Abducent nerve**: see p. 145.
• **Three divisions of the trigeminal nerve in the cavum trigeminale**: see Fig. 73.3.

The cerebrospinal fluid

The cerebrospinal fluid (CSF) is a clear fluid that surrounds the brain and spinal cord. It is secreted by the choroid plexuses of the lateral, 3rd and 4th ventricles of the brain and passes into the subarachnoid space of both the brain and the spinal cord through the median and lateral apertures in the roof of the 4th ventricle. It is absorbed by the arachnoid villi in the superior sagittal sinus.

Clinical notes

• **Lumbar puncture**: the spinal cord ends at the level of the lower border of L1, so that a needle may be passed into the subarachnoid space below this level without fear of damaging the cord. It is normally performed with the patient lying on his or her side with the lumbar spine fully flexed so as to open up the spaces between the spines of the vertebra. A long needle is then passed in the midline through or closely alongside the supraspinous and interspinous ligaments and then through the serous layer of the dura and the arachnoid. The fluid normally escapes drop by drop, but when it is under increased pressure it is in a continuous stream. The pressure may be measured by means of a manometer and is usually about 100 mm CSF, but may be in the range 50–180 mm CSF.

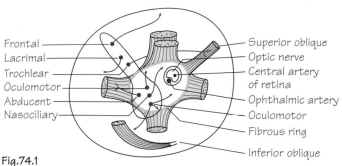

Fig.74.1

The back of the orbit to show the origins of the muscles that move the eyeball and the nerves that enter through the superior orbital fissure and the optic canal

Frontal
Lacrimal
Trochlear
Oculomotor
Abducent
Nasociliary

Superior oblique
Optic nerve
Central artery of retina
Ophthalmic artery
Oculomotor
Fibrous ring
Inferior oblique

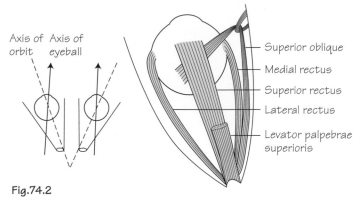

Fig.74.2

The muscles that move the eyeball seen from above

Axis of orbit   Axis of eyeball

Superior oblique
Medial rectus
Superior rectus
Lateral rectus
Levator palpebrae superioris

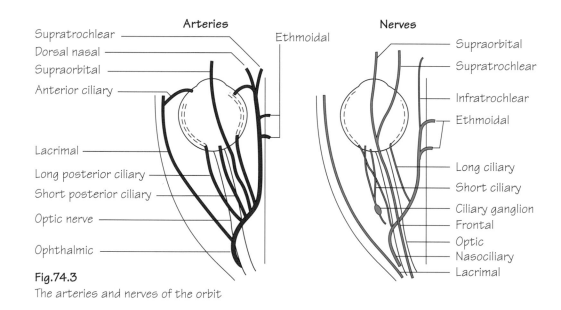

Arteries

Supratrochlear
Dorsal nasal
Supraorbital
Anterior ciliary

Ethmoidal

Lacrimal
Long posterior ciliary
Short posterior ciliary
Optic nerve
Ophthalmic

Nerves

Supraorbital
Supratrochlear
Infratrochlear
Ethmoidal

Long ciliary
Short ciliary
Ciliary ganglion
Frontal
Optic
Nasociliary
Lacrimal

Fig.74.3

The arteries and nerves of the orbit

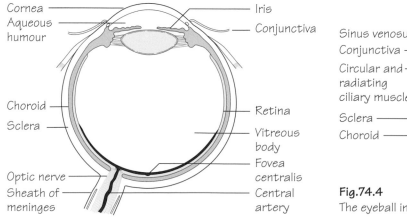

Cornea
Aqueous humour

Iris
Conjunctiva

Choroid
Sclera

Retina
Vitreous body
Fovea centralis

Optic nerve
Sheath of meninges

Central artery

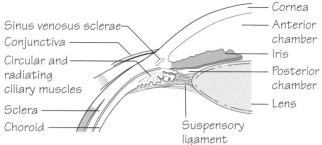

Sinus venosus sclerae
Conjunctiva
Circular and radiating ciliary muscles
Sclera
Choroid

Cornea
Anterior chamber
Iris
Posterior chamber
Lens
Suspensory ligament

Fig.74.4

The eyeball in section and detail of the iridocorneal angle

The bony walls of the orbit have already been described (Chapter 60).

The orbit contains the eyeball and optic nerve, the extraocular muscles, the lacrimal gland, the 3rd, 4th and 6th cranial nerves and the three branches of the ophthalmic division of the trigeminal nerve.

The parasympathetic *ciliary ganglion* is attached to a branch of the oculomotor nerve.

The *ophthalmic artery* supplies the contents of the orbit, and the *superior* and *inferior ophthalmic veins* drain it, passing through the superior orbital fissure into the cavernous sinus.

- **The superior orbital fissure**: this slit-like opening is divided into two parts by the fibrous ring that forms the origin of the main muscles that move the eyeball. The ring also includes the *optic canal* (Fig. 74.1). It is the portal of entry for a number of important structures:
  - *Above the ring*: frontal, lacrimal and trochlear nerves.
  - *Within the ring*: two divisions of the oculomotor, the nasociliary and the abducent nerves.
- **The optic canal**: transmits the optic nerve and the ophthalmic artery.
- **The inferior orbital fissure**: transmits the maxillary nerve and some small veins.
- **The muscles of the eyeball (Fig. 74.2)**: except for the inferior oblique, these originate from the fibrous ring and spread out to form a fibromuscular cone that encloses the eyeball. The ring also gives origin to the *levator palpebrae superioris* which is inserted into the upper eyelid and opens the eye.
  - The *lateral rectus*: turns the eyeball laterally.
  - The *medial rectus*: turns the eyeball medially.
  - The *superior rectus*: because of the different long axes of the orbit and eyeball, it turns the eye upwards and medially.
  - The *inferior rectus*: for the same reason mentioned above, it turns the eye downwards and medially.
  - The *superior oblique*: passes along the medial wall of the orbit, turns sharply through a fibrous pulley and is inserted into the upper part of the eyeball, below the superior rectus. It turns the eye downwards and laterally. When this muscle and the inferior rectus contract together, the eye turns directly downwards.
  - The *inferior oblique*: arises from the floor of the orbit, passes under the eyeball like a hammock and is inserted into its lateral side. It turns the eye upwards and laterally. Together with the superior rectus, *the inferior oblique* turns the eye directly upwards.
- **The nerve supply of the orbital muscles**: the lateral rectus ('abductor') of the eye is supplied by the *abducent nerve*; the superior oblique (the 'muscle with the pulley') is supplied by the *trochlear nerve*. All other muscles, including levator palpebrae superioris, are supplied by the *oculomotor nerve*.
- **The other nerves and vessels of the orbit** are shown in Fig. 74.3. The most important branch of the ophthalmic artery is the *central artery of the retina* which enters and passes through the optic nerve and is the only blood supply to the retina.

The eyeball (Fig. 74.4)

The eyeball is composed of three layers. The outermost is a tough fibrous layer, the *sclera*. Within this is the very vascular *choroid* and inside this again is the sensory part of the eye, the *retina*. Anteriorly, the sclera is replaced by the transparent *cornea*, which is devoid of vessels or lymphatics. At the corneoscleral junction, there is an important venous structure, the *sinus venosus sclerae (canal of Schlemm)*. Behind the cornea, the choroid is replaced by the *ciliary body*, with its radially arranged *ciliary processes*, and the *iris*. The ciliary body contains the circular and radial smooth muscle fibres of the *ciliary muscle*, supplied by parasympathetic fibres from the ciliary ganglion via the oculomotor nerve. These, when they contract, relax the lens capsule and allow the lens to expand; thus, they are used in near vision. The iris contains smooth muscle fibres of the *dilator pupillae* and *sphincter pupillae*, supplied, respectively, by the *sympathetic system* (from the *superior cervical ganglion*) and the *parasympathetic system* (from the *oculomotor nerve* via the *ciliary ganglion*). The *lens* lies behind the pupil and is enclosed in a delicate *capsule*. It is suspended from the ciliary processes by the *suspensory ligament*.

The ciliary body secretes the *aqueous humour* into the *posterior chamber* of the eye (lying behind the pupil). The aqueous then passes through the pupil into the *anterior chamber* and is reabsorbed into the sinus venosus sclerae.

Behind the lens, the eyeball is occupied by the gelatinous *vitreous humour*.

The *retina* consists of an inner nervous layer and an outer pigmented layer. The nervous layer has an innermost layer of ganglion cells whose axons pass back to form the optic nerve. Outside this is a layer of bipolar neurones and then the receptor layer of rods and cones. Near the posterior pole of the eye is the yellowish *macula lutea*, the receptor area for central vision. The *optic disc* is a circular pale area marking the end of the optic nerve and the site of entry of the central artery of the retina. This divides into upper and lower branches, each of which gives temporal and nasal branches.

Clinical notes

- **Glaucoma**: any blockage of the drainage of the aqueous into the sinus venosus sclerae will cause a rise in the intra-ocular pressure, a condition known as *glaucoma*. Unless early treatment is given, it can lead to deterioration in vision and eventual blindness.

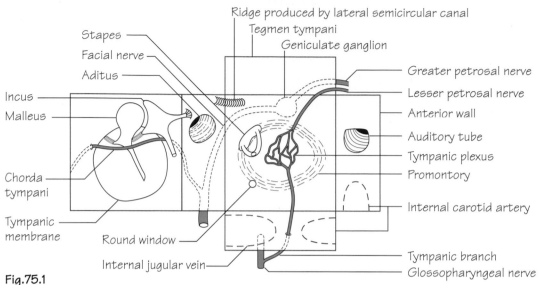

Fig.75.1

A diagram representing the right middle ear as an opened-out box

Labels (clockwise): Stapes, Facial nerve, Aditus, Incus, Malleus, Chorda tympani, Tympanic membrane, Round window, Internal jugular vein, Ridge produced by lateral semicircular canal, Tegmen tympani, Geniculate ganglion, Greater petrosal nerve, Lesser petrosal nerve, Anterior wall, Auditory tube, Tympanic plexus, Promontory, Internal carotid artery, Tympanic branch, Glossopharyngeal nerve

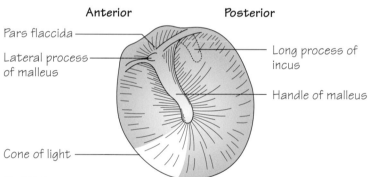

Anterior   Posterior

Pars flaccida
Lateral process of malleus
Long process of incus
Handle of malleus
Cone of light

Fig.75.2

The left tympanic membrane, as seen through an auriscope. The 'cone of light' is caused by the reflection of the light of the auriscope

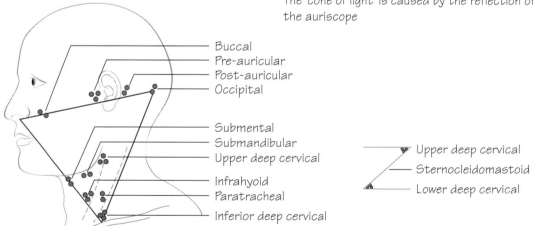

Buccal
Pre-auricular
Post-auricular
Occipital

Submental
Submandibular
Upper deep cervical
Infrahyoid
Paratracheal
Inferior deep cervical

Upper deep cervical
Sternocleidomastoid
Lower deep cervical

Fig.75.3

The principal groups of lymph nodes of the head and neck arranged as a triangle. The inset (right) shows the two major groups into which the others eventually drain

The ear

The ear is subdivided into the outer ear, the middle ear and the inner ear.

The outer ear

The outer third of this is cartilaginous and the inner two-thirds is bony. Lined by skin with ceruminous (wax) glands.

The middle ear

This has four walls, a roof and a floor. It can, therefore, be represented diagrammatically by an opened-out box (Fig. 75.1).

- The lateral wall:
  - The *tympanic membrane* (Fig. 75.2): the handle of the malleus is embedded in its middle layer. This is crossed by the *chorda tympani*, above which is the *pars flaccida*. The *cone of light* is the reflection of the light from the auriscope.
  - The *epitympanic recess (attic)*: is the part of the middle ear cavity above the tympanic membrane.
  - The *ossicles: the malleus, incus and stapes*: the stapes engages with the oval window. The ossicles transmit vibrations of the membrane to the inner ear.
- **The medial wall**:
  - The *promontory*: is a bulge produced by the first turn of the cochlea.
  - The *oval window*: leads into the inner ear.
  - The *facial nerve*: runs backwards and then downwards in a bony canal in the medial wall. It bears the geniculate ganglion and gives off the *chorda tympani*.
- **The anterior wall**: the *pharyngotympanic (Eustachian) tube* opens into the anterior wall and leads down to the nasopharynx. Its function is to equalise the pressure between the middle ear and the pharynx, and therefore the atmosphere.
- **The posterior wall**: the *aditus* leads backwards into the *mastoid antrum*, a cavity in the mastoid bone, which, in turn, leads into the *mastoid air cells*.
- **The roof**: the *tegmen tympani* is a thin plate of bone that separates the middle ear from the middle cranial fossa.
- **The floor**: separates the middle ear from the internal carotid artery and the internal jugular vein.

The inner ear

The inner ear is involved in both hearing and balance. It consists of two components:

- **The osseous labyrinth**: comprises the *vestibule*, the *semicircular canals* and the *cochlea*. The labyrinth itself consists of spaces in the petrous temporal bone and contains the *membranous labyrinth*.
- **The membranous labyrinth**: comprises the *utricle* and *saccule* (in the vestibule), the *semicircular ducts* (in the semicircular canals) and the *duct of the cochlea* (in the cochlea). The utricle and saccule are concerned with the sense of position and the semicircular ducts are concerned with the sensation of motion. The cochlear duct is the organ of hearing. All are supplied by the *vestibulocochlear (auditory) nerve*.

The lymphatics of the head and neck
(Fig. 75.3)

- **Upper deep cervical nodes (*jugulodigastric nodes*)**: these are situated between the upper end of sternomastoid and the angle of the mandible and also deep to sternomastoid. They drain the head and the upper part of the neck, directly or indirectly (but there are no lymphatics in the cranial cavity).
- **Lower deep cervical nodes (*jugulo-omohyoid nodes*)**: these are situated in the posterior triangle between the lower end of sternomastoid and the clavicle. They drain the lower part of the neck and also receive lymph from the upper deep cervical nodes, from the breast and some of the lymph from the thorax and abdomen. The efferents from this group drain into the thoracic duct or the right lymph duct via the *jugular trunk*.
- **Smaller groups of nodes** are shown in Fig. 75.3.
- **The lymph drainage of the tongue**: the tip of the tongue drains into the *submental nodes*. The rest of the anterior two-thirds drains into the *submandibular nodes*, some crossing the midline and some passing directly to the *upper deep cervical nodes*. The posterior one-third drains directly into the *upper deep cervical nodes*.
- **The lymph drainage of the larynx**: above the vocal cords, the larynx drains into the *infrahyoid nodes* and thence to the *upper deep cervical group*. Below the cords, drainage is to the *paratracheal and inferior deep cervical nodes*.

Surface anatomy of the head and neck

- **The middle meningeal artery**: the *anterior branch* may be exposed at a point 4 cm above the midpoint of the zygomatic arch. The *posterior branch* may be represented on the surface by a pencil placed behind the ear.

The face

- **The supraorbital, infraorbital and mental nerves,** all lie on a vertical line passing between the two premolar teeth.
- **The facial artery** can be felt on the mandible at the anterior border of the masseter.
- **The superficial temporal artery** is just in front of the tragus of the ear.
- **The parotid duct** follows the middle part of a line from the tragus of the ear to the middle of the upper lip. It hooks over the anterior border of the masseter where it can be easily felt when the teeth are clenched.

The neck

- **The sternomastoid muscle** (with the *external jugular vein* on its surface) may be made to contract by asking the patient to turn his or her head to the *opposite* side against resistance.
- **The trunks of the brachial plexus** can be palpated in the angle between the sternomastoid and the clavicle.
- **The subclavian artery** is palpable by deep pressure behind the middle of the clavicle. A cervical rib may also be palpable.
- **The hyoid bone, and the thyroid and cricoid cartilages** are easily felt. The larynx, and any swellings associated with it, moves upwards on swallowing, as does the thyroid gland. *Thyroglossal cysts* move upwards when the tongue is protruded (p. 173).
- **The trachea** is palpable in the suprasternal notch.
- **The common carotid artery** can be felt in front of sternomastoid and can be compressed against the transverse process of the 6th cervical vertebra (*carotid tubercle*).

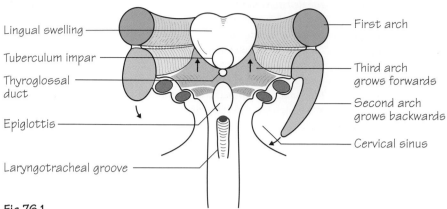

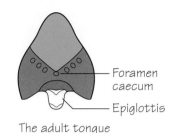

Fig.76.1
The floor of the pharynx to show the development of the tongue

The adult tongue

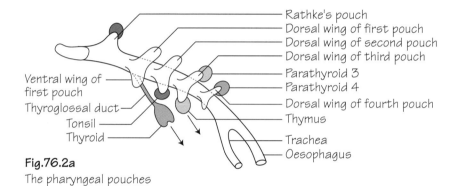

Fig.76.2a
The pharyngeal pouches

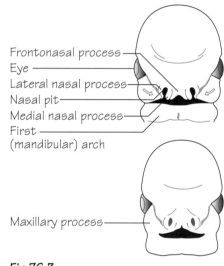

Fig.76.3
Development of the face

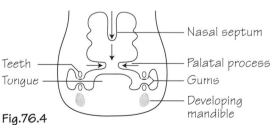

Fig.76.2b
The fate of the pharyngeal pouch components seen in Fig 76.2a

Fig.76.4
Coronal section through developing mouth and nasal cavity

The pharyngeal region, development

There are, theoretically, six *branchial arches* in this region, but the 5th arch never develops in the human embryo. Each arch is, in fact, an inverted U-shaped arch lying in the floor and lateral walls of the pharynx (Fig. 76.1). Each consists of a core of *mesoderm* covered on the outside by *ectoderm* and on the inside by *endoderm*. The latter projects laterally between each pair of arches to form the *pharyngeal pouches*, each of which has *dorsal* and *ventral wings* (Fig. 76.2). Cranial nerves grow down into the arches, and these are as follows: 1st arch, *trigeminal*; 2nd arch, *facial*; 3rd arch, *glossopharyngeal*; 4th arch, *superior laryngeal branch of the vagus*; 6th arch, *recurrent laryngeal branch of the vagus*.

- **The cervical sinus**: the large 2nd arch grows backwards to cover the remaining smaller arches and fuses with the region behind the 6th arch to form a closed epithelial-lined cavity, *the cervical sinus* (Fig. 76.1). This normally disappears.
- **The derivatives of the pharyngeal pouches (Fig. 76.2)**: the first and dorsal wings of the 2nd pouch elongate to form the middle ear cavity and the Eustachian tube. The endoderm of the remaining pouches proliferates to form:
  - *The tonsils*: develop from the ventral wing of the 2nd pouch which becomes infiltrated with lymphocytes.
  - *The thymus*: develops from the endoderm of the ventral wing of the 3rd pouch and, together with the heart and great vessels, descends into the thorax, the two sides fusing to form a single gland. It loses its connection with the pharynx and becomes infiltrated with thymocytes.
  - *The parathyroid glands*: from the 3rd and 4th pouches, and are best called parathyroids 3 and 4 because their relative positions change. Parathyroid 3 is closely associated with the thymus and descends with it until it becomes embedded in the back of the thyroid to form the *inferior parathyroid*. Parathyroid 4 descends as far as the upper part of the thyroid to form the *superior parathyroid*.
- **The thyroid gland**: an outgrowth, the *thyroglossal duct*, develops from the floor of the pharynx just behind the tuberculum impar and descends ventral to the developing hyoid bone. It divides into two lobes and loses its connection with the floor of the pharynx when the thyroglossal duct disappears.
- **The tongue**: this develops from two *lingual swellings* on the 1st arch (Fig. 76.1) which take in the *tuberculum impar* and form the anterior two-thirds of the tongue. The site of the thyroglossal outgrowth is marked by the *foramen caecum*. The posterior two-thirds is formed from the *3rd arch mesoderm* which grows forwards over the 2nd arch. The 4th arch makes a small contribution to the back of the tongue. The musculature comes from the *occipital mesoderm somites* and is, therefore, supplied by the *hypoglossal nerve*.
- **The pituitary gland**: a pharyngeal upgrowth, Rathke's pouch, forms the anterior lobe, while the posterior lobe develops from a downgrowth from the brain (Fig. 76.2).

The pharyngeal region, developmental anomalies

- **Branchial cyst**: the cervical sinus may not disappear ,and this is thought to be the cause of a *branchial cyst* which presents as a swelling at the anterior border of sternomastoid, containing a cholesterol-rich fluid. If the sinus remains open to the exterior, it results in a *branchial fistula*, a small opening in the same situation which usually becomes infected and has to be removed. It leads upwards between the internal and external carotid arteries and may even open into the tonsillar bed.
- **Aberrant parathyroid**: the inferior parathyroid may retain its association with the thymus, so that it may be found lower down in the neck than usual and may even be found in the chest.
- **Thyroglossal remnants**: the thyroglossal duct may persist and thyroglossal cysts or aberrant thyroid tissue may develop along its course. There may be a nodule of thyroid tissue on the tongue at the foramen caecum (*lingual thyroid*), or there may be a swelling lower in the neck. In such cases, the swelling will move upwards on protruding the tongue.

The face and palate, development

In the future face region, a pair of *nasal sacs* appears, the openings of which form the *external nares*, while internally they open into the primitive mouth cavity (*stomatodaeum*). On either side of each, there are *medial* and *lateral nasal processes* (Fig. 76.3) and the medial processes, together with the area between them, form the *frontonasal process*. The *maxillary process* grows forwards below the developing eye to meet the lateral nasal process. Below the mouth, the lower jaw is forming from the 1st branchial (mandibular) arch. The maxillary process continues to grow medially and reaches and fuses with the frontonasal process, thus completing the upper lip, the line of fusion between them running up to the inner canthus of the eye, which faces laterally at this stage. Within the stomatodaeum, a pair of *palatal processes* grows medially and a *nasal septum* grows down from the roof. All three processes unite with each other and with the inner aspect of the frontonasal process to separate the nasal and mouth cavities (Fig. 76.4). Later, bone and muscle will develop to form the hard and soft palates.

The face and palate, developmental anomalies

- **Cleft lip and palate**: failure of the maxillary process to fuse with the frontonasal process results in a *cleft lip* in which there is a unilateral fissure in the lip. It may be repaired surgically, usually at about 3 months of age. If one of the palatal processes fails to reach the midline and to fuse with the frontonasal process, there will be a unilateral cleft palate which often leads into a cleft lip on the same side. The least severe form of this anomaly is a bifid uvula. Bilateral clefts lead to a more serious deformity as there is a wide cleft in the palate through which the lower edge of the septum can be seen. The free frontonasal process projects forwards as a distorted swelling. Cleft palate gives rise to difficulties with feeding as it is not possible to produce the negative intra-oral pressure which is necessary for sucking, and also because the mouth cannot be closed off from the nasal cavity so that regurgitation of milk occurs. Later, there are problems with speech and with ear infections owing to interference with the palatal muscles that are attached to the Eustachian tube. Surgical treatment, which is very effective, is usually carried out between 9 and 12 months of age.

77 The spine

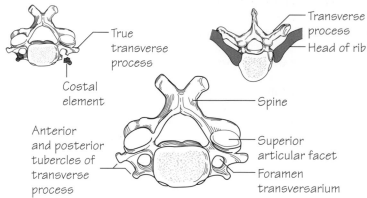

Fig.77.1
A cervical vertebra.
The anterior tubercles of the transverse processes represent the costal elements

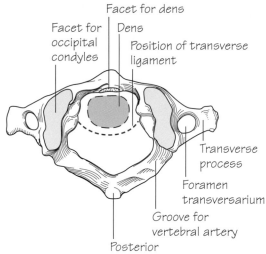

Fig.77.2
The upper surface of the atlas.
The dotted area shows the position of the dens and the dotted line indicates the transverse ligament

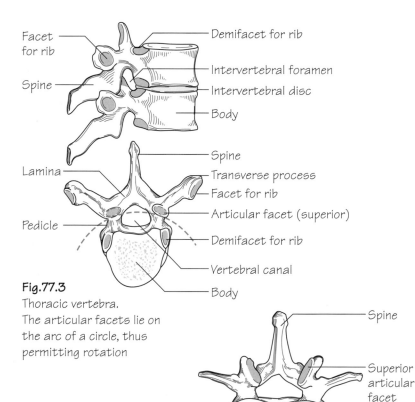

Fig.77.3
Thoracic vertebra.
The articular facets lie on the arc of a circle, thus permitting rotation

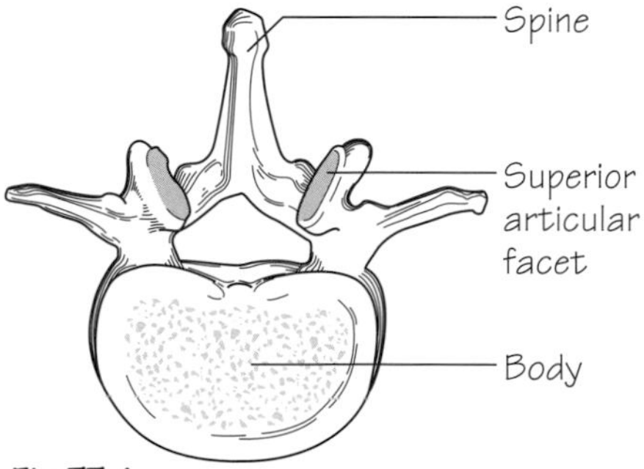

Fig.77.4
A lumbar vertebra.
The direction of the articular facets does not permit rotation

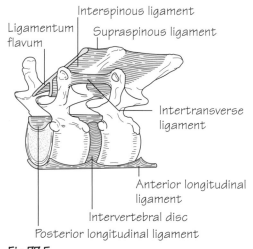

Fig.77.5
The ligaments joining two adjacent vertebrae

 Anatomy at a Glance, Third Edition. Omar Faiz, Simon Blackburn and David Moffat.

There are 7 cervical, 12 thoracic, 5 lumbar, 5 sacral and 3–5 coccygeal vertebrae. They are held together by ligaments, intervertebral discs and synovial joints between the articular processes. The weight-bearing portion of the vertebra is the body, so that the size of the vertebral bodies increases from above downwards. The bodies contain red bone marrow, so that the veins that drain them (*basivertebral veins*) are large and occupy large foramina on the backs of the bodies.

- **Cervical vertebrae (Fig. 77.1)**: have small *bodies*, *bifid spines*, *transverse processes* with *anterior* and *posterior tubercles,* and a *foramen transversarium* for the passage of the vertebral artery. The anterior tubercles represent the *costal element* (Fig. 77.1) and may be enlarged to produce a cervical rib (Chapter 3). The body of the first vertebra (*atlas*) fuses with that of the second (*axis*) during development to produce its *dens* (*odontoid process*), which is held in place by a transverse ligament (Fig. 77.2).
- **Thoracic vertebrae (Fig. 77.3)**: have heart-shaped *bodies*, upper and lower *demifacets* for the heads of the ribs, long downturned *spine*, and long *transverse processes* with a *facet* for the tubercles of the rib.
- **Lumbar vertebrae (Fig. 77.4)**: have a massive *body*, large *transverse processes*, a triangular *vertebral canal,* and large, backwardly projecting *spines* so that a needle may be inserted between them in the procedure of *lumbar puncture* (Chapter 73). The *articular facets* face mediolaterally to prevent rotation.
- **Sacral vertebrae (Fig. 27.3)**: five vertebrae are fused together to form the *sacrum*.

The intervertebral joints

The upper and lower surfaces of the bodies are covered with hyaline cartilage and are separated by the fibrocartilaginous *intervertebral discs*. Each disc has a peripheral fibrous ring (*annulus fibrosus*) and a central, more spongy *nucleus pulposus* which lies nearer to the back than to the front of the disc. The nucleus pulposus is rich in glycosaminoglycans, so that it has high water content, but this diminishes with increasing age. The discs are thickest in the lumbar and cervical regions, so that these are the regions with the greatest range of movement.

The vertebrae are also held together by ligaments that join each of the components of the vertebrae (Fig. 77.5) except for the pedicles (the spinal nerves have to pass between these in the intervertebral foramina). Thus, there are:

- **Anterior and posterior longitudinal ligaments** joining the front and back of the bodies (the posterior ligaments, thus, lie within the vertebral canal).
- **Supraspinous and interspinous ligaments.**
- **Intertransverse ligaments.**
- **Ligamenta flava** (which contain much elastic tissue and join the laminae).
- **Capsular ligaments** of the synovial joints between the articular processes.

Curves of the spine

In the fetus, the whole developing spine is curved, so that it is concave forwards (*primary curvature*). A few months after birth, the baby begins to hold its head up and the cervical spine develops a forward convexity (*secondary curvature*). Later in the first year, the baby begins to stand up and another forward convexity develops in the lumbar region (*secondary curvature*). The primary curves are mainly due to the shape of the vertebral bodies, but the secondary curves are due to the shape of the intervertebral discs.

Movements of the spine

- **Cervical spine**: movement is free in the cervical region of the spine: *flexion and extension*, *side flexion* (associated with rotation to the opposite side) and *rotation*. In rotation of the head, the atlas rotates around the dens of the axis, and in flexion of the head, the occipital condyles move on the articular facets of the atlas.
- **Thoracic spine**: movement is somewhat restricted by the thinner intervertebral discs and the presence of the ribs. *Rotation*, however, is free.
- **Lumbar spine**: *flexion and extension* are free but *rotation* is almost absent because of the direction of the articular facets. Side flexion also occurs.

Clinical notes

- **Spina bifida**: each vertebra develops in three parts – the body and the two halves of the neural arch. These ossify in cartilage and, at birth, the three parts are still separated by cartilage, but they soon fuse. Failure of such fusion gives rise to the condition of *spina bifida*. This may be symptomless (*spina bifida occulta*), although the site of the lesion may be marked by a tuft of hair on the surface. If the defect is large, the meninges, or even the spinal cord, may herniate onto the surface (*meningocoele*), and this may produce neurological symptoms.
- **Prolapsed intervertebral disc**: the nucleus pulposus may herniate through the annulus fibrosus, which is thinnest posteriorly, passing *backwards* (compressing the spinal cord), *posterolaterally* (compressing a spinal nerve) or *upwards* into a vertebral body – a *Schmorl's node*.
- **Fracture of the spine**: this is most commonly the result of acute flexion of the spine, so that the bodies of the vertebrae are compressed. Such an injury can occur in the cervical region as a result of diving into shallow water or in the thoracic and lumbar regions by falling from a height or by heavy objects falling on the trunk. When severe, it may be accompanied by dislocation and damage to the spinal cord or cauda equina.
- **Disease of the vertebral bodies**: as the bodies of the vertebrae are responsible for weight-bearing, disease of the bodies, such as osteoporosis or tuberculous, can cause collapse of one or more bodies with consequent deformity.

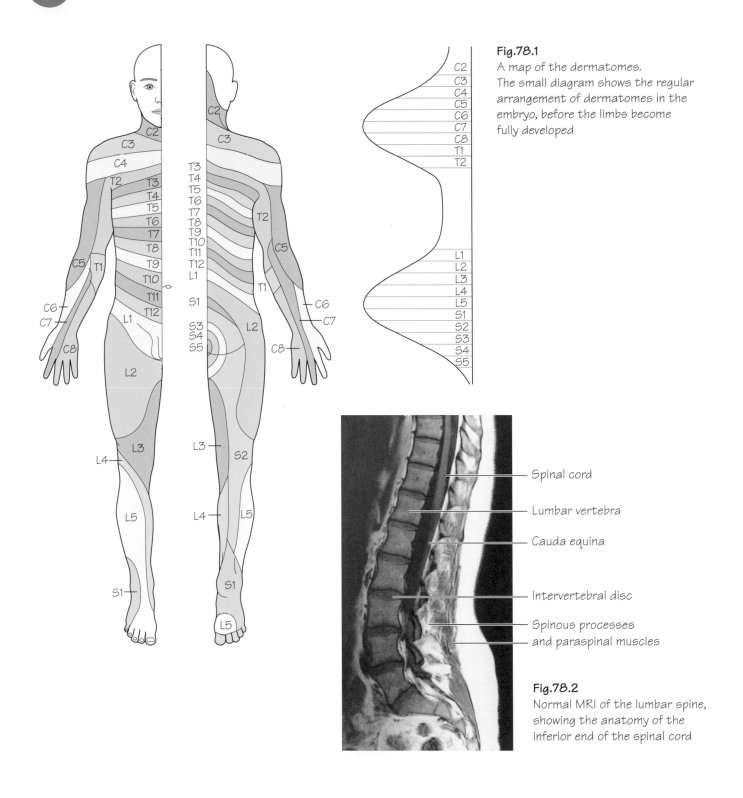

Fig.78.1
A map of the dermatomes.
The small diagram shows the regular arrangement of dermatomes in the embryo, before the limbs become fully developed

Spinal cord

Lumbar vertebra

Cauda equina

Intervertebral disc

Spinous processes
and paraspinal muscles

Fig.78.2
Normal MRI of the lumbar spine, showing the anatomy of the inferior end of the spinal cord

In the fetus, the spine and the spinal cord are the same length but, as the spine grows more quickly than the spinal cord, the lower end of the cord gradually retreats upwards, reaching the level of L3 at birth and the lower border of L1 in the adult (Fig. 78.2). For this reason, too, the anterior and posterior nerve roots become more and more oblique from above downwards, so the lumbar and sacral nerve roots form a bundle, the *cauda equina*, which occupies the lower part of the spinal canal. The spinal cord itself ends as the *filum terminale*, a thin fibrous band which is included in the cauda equina.

The spinal cord shows two enlargements in the cervical and lumbar regions, corresponding to the origins of the nerves that make up the limb plexuses. It is for this reason that the vertebral canal is larger in these regions, and they are also the regions of the greatest mobility.

• **The meninges**: the meninges – the *dura* (serous layer only), *pia* and *arachnoid* – are continuous with those of the cranial cavity, so that the *subarachnoid space* is also continuous and cerebrospinal fluid may be drained from the system or its pressure measured, by *lumbar puncture* (Chapter 73). The subarachnoid space ends at the level of S2.

• **The blood supply of the spinal cord**: the spinal cord is supplied with blood by *spinal arteries* that are derived from the *vertebral*, *intercostal*, *lumbar* and *lateral sacral arteries*. Each artery divides into *dorsal* and *ventral* branches that follow the corresponding nerve roots to the spinal cord where they form longitudinally arranged *anterior* and paired *posterior spinal arteries*. These longitudinal anastomoses, however, are incomplete and the spinal arteries themselves vary in size. The largest are in the lower thoracic and upper lumbar regions, and the blood supply of the cord may be jeopardised if some of these larger spinal arteries are damaged, for example, in resection of the thoracic aorta.

The spinal nerves

Each spinal nerve arises by *anterior* and *posterior* (ventral and dorsal) *nerve roots*. Each posterior root bears the *posterior root ganglion*. Each nerve cell of the ganglia has one process which divides into two – a central process that enters the spinal cord and a peripheral process that is the axon of a peripheral nerve. Thus, there are no synapses in the ganglion. The anterior nerve roots carry *motor fibres* arising from the anterior horn cells and the posterior nerve roots carry *sensory fibres* from the cells of the posterior root ganglia. The posterior root ganglia lie in the intervertebral foramina, so that the lower nerve roots have a long course in the cauda equina before they unite to form spinal nerves.

As described in Chapter 4, the spinal nerves from T2 to L1 supply segmental areas of skin and muscles, but elsewhere they form plexuses, so that the areas supplied by each nerve become more complicated. Each named *peripheral nerve* may contain components of more than one *spinal nerve* and each *spinal nerve* may contribute fibres to more than one *peripheral nerve*. However, the area of skin supplied by any one spinal nerve (a *dermatome*) can be mapped out, and this is shown in Fig. 78.1. In the embryo, the area supplied by the spinal nerve follows a regular segmental pattern (Fig. 78.1), but with the later development of the limbs, which involves rotation, the map becomes more complicated. Knowledge of the dermatome map is of great value in diagnosing lesions of the spinal nerves or of segments of the spinal cord, but it must be remembered that these areas are subject to some variation from one person to another and there is also a good deal of overlap. Thus, a lesion of, say, T4 may cause little or no loss of sensation because of overlap with T3 and T5.

Clinical notes

• **Herpes zoster**: individual dermatomes can be visualized in a virus infection of one or more posterior root ganglia, called *herpes zoster*, which gives rise to a rash over the corresponding dermatome. It normally resolves spontaneously, but is sometimes followed by a painful *post-herpetic neuralgia* (see also Chapter 62).

• **Myotomes**: certain spinal nerves supply well-defined groups of muscles, sometimes called *myotomes*. For example, C5 supplies the *abductors of the shoulder*, the *flexors of the elbow* and the *supinators*, so that a lesion of this nerve or of the upper trunk of the brachial plexus will cause the arm to be held in a position of adduction of the shoulder, extension of the elbow and pronation, sometimes called the 'waiter's tip position' (*Erb–Duchenne paralysis*). Similarly, T1 supplies the small muscles of the hand, so that a lesion of this nerve, due perhaps to the presence of a cervical rib, will cause global wasting of the hand muscles.

• **Transection of the cord**: road traffic accidents are the most common cause of spinal cord injuries. If the cord is severely damaged, there will be a loss of all sensation below the level of the lesion and paralysis of all movement. The paralysis is at first flaccid, but becomes spastic at a later stage. In high cervical lesions, there is paralysis of the intercostal muscles and also the diaphragm (phrenic nerve, C3, 4 and 5), so that ventilation is needed.

Muscle index

The abdomen

All the muscles of the *anterior abdominal wall* serve to protect the viscera by their contraction, to produce movement and to increase the intra-abdominal pressure, as in defecation, coughing, parturition, etc. All these muscles are supplied by the *lower six thoracic* and the *first lumbar nerves*.

- **External oblique**

Origin: From the outer surfaces of the lower eight ribs to the iliac crest as far forwards as the anterior superior spine.

Insertion: The free lower border forms the *inguinal ligament* between the anterior superior iliac spine and the pubic tubercle. The muscle becomes aponeurotic and reaches the midline where it interdigitates with the opposite side to form the *linea alba*. The *superficial inguinal ring* is a gap in the aponeurosis above and medial to the pubic tubercle. The aponeurosis contributes to the *anterior rectus sheath*.

Actions: Flexion of the spine, side flexion and rotation of the trunk – the *right* external oblique produces rotation to the *left*.

- **Internal oblique**

Origin: From the thoracolumbar fascia, the iliac crest and the lateral half of the inguinal ligament.

Insertion: Into an aponeurosis which is attached to the costal margin and to the linea alba, after splitting to enclose the rectus abdominis, thus contributing to the rectus sheath. The lower fibres from the inguinal ligament contribute to the *conjoint tendon* which is attached to the pubic crest and the pectineal line.

Actions: Flexion of the spine, side flexion and rotation of the trunk – the *right* internal oblique produces rotation to the *right*.

- **Transversus abdominis**

Origin: From the thoracolumbar fascia, the iliac crest and the lateral one-third of the inguinal ligament. Also from the *inner* surfaces of the lower six ribs, interdigitating with the diaphragm.

Insertion: The aponeurosis passes to the linea alba, contributing to the rectus sheath. The lower fibres help to form the conjoint tendon.

Actions: Most of the fibres are transverse and thus pull in and flatten the abdominal wall.

- **Rectus abdominis**

Origin: From the anterior surfaces of the fifth, sixth and seventh costal cartilages.

Insertion: The pubic crest and tubercle and the front of the symphysis. There are three tendinous intersections in the upper part of the muscle which are adherent to the anterior rectus sheath.

Actions: The muscle is enclosed in the rectus sheath. A strong flexor of the trunk, it can also tilt the pelvis backwards. In a person lying prone, rectus contracts when the head is lifted from the pillow or when the leg is raised from the bed.

- For further important details of the muscles of the anterior abdominal wall, see the *inguinal canal* and the *rectus sheath* (p. 40).

- **The diaphragm**

Origin: From the inner surfaces of the lower six ribs, from the back of the xiphisternum, from the right and left crura which are attached, respectively, to the upper three and upper two lumbar vertebrae, and from the medial and lateral arcuate ligaments which bridge over the psoas major and quadratus lumborum.

Insertion: The fibres (striated muscle) are inserted into the central tendon.

Actions: The diaphragm is involved in respiration. When the muscle fibres contract, the diaphragm is lowered, thus increasing the vertical dimension of the thorax. In the later stages of contraction, using the liver as a fulcrum, it raises the lower ribs, thus increasing the width of the lower thorax. At the same time, it increases the intra-abdominal pressure and is thus used in expulsive efforts – defecation, micturition, parturition, etc.

Nerve supply: Phrenic nerve (C3, 4 and 5).

- **Quadratus lumborum**

Origin: From the posterior part of the iliac crest.

Insertion: To the twelfth rib.

Action: Side flexion of the trunk.

Nerve supply: Adjacent lumbar nerves.

- **Psoas major:** see lower limb (p. 181).

The upper limb

- **Latissimus dorsi**

Origin: From the spines of the lower six thoracic vertebrae, the lumbar spines via the thoracolumbar fascia and the medial part of the iliac crest, together with a small number of fibres from the inferior angle of the scapula.

Insertion: To the floor of the intertubercular sulcus, curving round teres major.

Actions: Adduction and medial rotation of the arm. Can hold up the lower limb girdle as in crutch walking.

Nerve supply: Thoracodorsal nerve.

- **Serratus anterior**

Origin: From the lateral surfaces of the upper eight ribs.

Insertion: Into the medial border of the scapula.

Actions: Protraction of the scapula and rotation, so that the glenoid points upwards, thus helping in abduction of the upper limb. Helps to keep the scapula in contact with the chest wall.

Nerve supply: Long thoracic nerve.

- **Levator scapulae**

Origin: From the transverse processes of the upper cervical vertebrae.

Insertion: To the medial border of the scapula above the spine.

Action: Elevates the scapula.

Nerve supply: C3 and 4.

- **The rhomboids**

Origin: From spines of thoracic vertebrae.

Insertion: To the medial border of the scapula.

Action: Bracing back the scapula.

Nerve supply: Dorsal scapular nerve.

- **Trapezius:** See p. 183.

- **Pectoralis major**

Origin: From the sternum and the upper six costal cartilages and from the medial half of the clavicle.

Insertion: To the lateral lip of the intertubercular sulcus.

Actions: Adduction, flexion and medial rotation of the arm.

Nerve supply: Medial and lateral pectoral nerves.

- **Pectoralis minor**

Origin: From the third, fourth and fifth ribs.

Insertion: To the coracoid process.

Action: Depresses the tip of the shoulder.

Nerve supply: Medial and lateral pectoral nerves.

- **Deltoid**

Origin: From the lateral third of the clavicle, the acromion and the spine of the scapula.

Insertion: To the deltoid tubercle on the lateral surface of the humerus.

Actions: Abduction of the upper limb (assisted by supraspinatus and serratus anterior), flexion (anterior fibres) and extension (posterior fibres) of the arm.

Nerve supply: Axillary nerve.

- **Teres major**

Origin: From the lower angle of the scapula.

Insertion: To the medial lip of the intertubercular sulcus.

Actions: Adduction and medial rotation of the arm.

Nerve supply: Lower subscapular nerve.

- **The rotator cuff**

Consists of *subscapularis*, *supraspinatus*, *infraspinatus* and *teres minor*. Acting together, these muscles maintain the stability of the shoulder joint as well as having their own individual actions, as follows:

- **Subscapularis**

Origin: From the subscapular fossa.

Insertion: Passes in front of the shoulder joint to the lesser tuberosity of the humerus.

Action: Medial rotation of the arm.

Nerve supply: Subscapular nerves.

- **Supraspinatus**

Origin: From the supraspinous fossa.

Insertion: To the top of the greater tuberosity of the humerus.

Action: Initiates abduction of the arm.

Nerve supply: Suprascapular nerve.

- **Infraspinatus**

Origin: From the infraspinous fossa.

Insertion: To the back of the greater tuberosity.

Action: Lateral rotation of the arm.

Nerve supply: Suprascapular nerve.

- **Teres minor**

Origin: From the lateral border of the scapula.

Insertion: To the humerus below infraspinatus.

Action: Lateral rotation of the arm.

Nerve supply: Axillary nerve.

- **Coracobrachialis**

Origin: From the coracoid process along with the short head of biceps and pectoralis minor.

Insertion: To the medial side of the humerus.

Action: Moves the arm upwards and medially.

Nerve supply: Musculocutaneous nerve.

- **Biceps brachii**

Origin: Long head from the supraglenoid tubercle and short head from the coracoid process (with coracobrachialis).

Insertion: To the radial tuberosity and, via the bicipital aponeurosis, into the deep fascia of the forearm.

Actions: Flexion and supination of the forearm.

Nerve supply: Musculocutaneous nerve.

- **Brachialis**

Origin: From the front of the lower part of the humerus.

Insertion: To the tubercle on the ulna just below the coronoid process.

Action: Flexion of the elbow.

Nerve supply: Musculocutaneous nerve and radial nerve.

- **Triceps**

Origin: Three heads: long from the infraglenoid tubercle, lateral from the humerus above the spiral line and medial from the back of the lower part of the humerus.

Insertion: Into the olecranon.

Action: Extensor of the elbow.

Nerve supply: Radial nerve.

- **Pronator teres**

Origin: From the common flexor origin on the medial epicondyle of the humerus.

Insertion: To the lateral surface of the shaft of the radius.

Action: Pronation of the forearm.

Nerve supply: Median nerve.

- **Flexor carpi radialis**

Origin: From the common flexor origin.

Insertion: To the base of metacarpals 2 and 3.

Actions: Flexion and abduction of the wrist.

Nerve supply: Median nerve.

- **Palmaris longus**

Origin: From the common flexor origin.

Insertion: To the flexor retinaculum and the palmar aponeurosis.

Action: Flexor of the wrist. (Often absent.)

Nerve supply: Median nerve.

- **Flexor carpi ulnaris**

Origin: From the common flexor origin and the posterior border of the ulna.

Insertion: To the pisiform and thence by the pisometacarpal ligament to the fifth metacarpal.

Actions: Flexion and adduction of the wrist.

Nerve supply: Ulnar nerve.

- **Flexor digitorum superficialis**

Origin: From the common flexor origin and the shaft of the radius.

Insertion: To the sides of the middle phalanges of the four fingers. The tendons are perforated by the tendons of flexor digitorum profundus.

Actions: Flexion of the proximal two phalanges and of the wrist.

Nerve supply: Median nerve.

- **Flexor pollicis longus**

Origin: From the front of the shaft of the radius.

Insertion: To the distal phalanx of the thumb.

Actions: Flexion of all the joints of the thumb.

Nerve supply: Median nerve.

- **Flexor digitorum profundus**

Origin: From the front of the shaft of the ulna and its posterior border.

Insertion: To the distal phalanges of the four fingers, the tendons passing through those of flexor digitorum superficialis.

Actions: Flexion of the fingers and the wrist.

Nerve supply: The lateral half by the median nerve and medial half by the ulnar nerve.

- **Pronator quadratus**

Origin: From the lower end of the front of the radius.

Insertion: To the lower end of the ulna.

Action: Pronator of the forearm.

Nerve supply: Median nerve.

- **Brachioradialis**

Origin: From the lateral supracondylar ridge of the humerus.

Insertion: To the lower end of the radius.

Action: Flexion of the elbow.

Nerve supply: Radial nerve.

- **Extensor carpi radialis longus and brevis**

Origin: From the lateral supracondylar ridge of the humerus.

Insertion: To the bases of the second and third metacarpals.

Actions: Extension and abduction of the wrist.

Nerve supply: Radial nerve.

- **Extensor digitorum**

Origin: From the common extensor origin on the lateral epicondyle of the humerus.

Insertion: To the bases of the middle and distal phalanges of the four fingers via the extensor expansion.

Actions: Extension of the fingers and of the wrist (but see also the lumbricals and interossei).

Nerve supply: Radial (posterior interosseous) nerve.

- **Extensor digiti minimi**

Origin: From the common extensor origin.

Insertion: To the extensor expansion of the little finger.

Action: Extension of the little finger.

Nerve supply: Radial (posterior interosseous) nerve.

- **Extensor carpi ulnaris**

Origin: From the common extensor origin and from the posterior border of the ulna.

Insertion: To the base of the fifth metacarpal.

Actions: Extension and adduction of the wrist.

Nerve supply: Radial (posterior interosseous) nerve.

- **Supinator**

Origin: From the lateral side of the humerus and the ulna.

Insertion: It wraps around the radius posteriorly to be inserted into the upper part of its shaft.

Action: Supination.

Nerve supply: Radial (posterior interosseous) nerve.

- **Abductor pollicis longus**

Origin: From the posterior surfaces of the radius and ulna.

Insertion: To the base of the first metacarpal.

Actions: Abduction and extension of the thumb.

Nerve supply: Radial (posterior interosseous) nerve.

- **Extensor pollicis brevis**

Origin: From the back of the radius.

Insertion: To the base of the proximal phalanx of the thumb.

Action: Extension of the proximal phalanx of the thumb.

Nerve supply: Radial (posterior interosseous) nerve.

- **Extensor pollicis longus**

Origin: From the back of the ulna.

Insertion: Into the base of the distal phalanx of the thumb.

Actions: Extension of all the joints of the thumb.

Nerve supply: Radial (posterior interosseous) nerve.

- **Extensor indicis**

Origin: From the back of the ulna.

Insertion: To the side of the extensor digitorum tendon to the index finger.

Action: Helps to extend the index finger.

Nerve supply: Radial (posterior interosseous) nerve.

- **Abductor pollicis brevis**

Origin: From the flexor retinaculum and adjacent carpal bones.

Insertion: To the base of the proximal phalanx of the thumb.

Action: Abduction of the thumb.

Nerve supply: Median nerve.

- **Flexor pollicis brevis**

Origin: From the flexor retinaculum and adjacent carpal bones.

Insertion: To the base of the proximal phalanx of the thumb. Its tendon contains a sesamoid bone.

Action: Flexion of the proximal phalanx of the thumb.

Nerve supply: Median nerve.

- **Opponens pollicis**

Origin: From the flexor retinaculum and adjacent carpal bones.

Insertion: To the shaft of the first metacarpal.

Action: Produces opposition of the thumb.

Nerve supply: Median nerve.

- **Adductor pollicis**

Origin: *Oblique head* from the bases of metacarpals; *transverse head* from the shaft of the third metacarpal.

Insertion: To the base of the medial side of the proximal phalanx of the thumb. The tendon contains a sesamoid bone.

Action: Adduction of the thumb.

Nerve supply: Deep branch of the ulnar nerve.

- **Abductor digiti minimi**

Origin: From the pisiform bone.

Insertion: To the base of the proximal phalanx of the little finger.

Action: Abduction of the little finger.

Nerve supply: Ulnar nerve.

- **Flexor digiti minimi**

Origin: From the flexor retinaculum and adjacent carpal bones.

Insertion: To the base of the proximal phalanx of the little finger.

Action: Flexes the proximal phalanx of the little finger.

Nerve supply: Ulnar nerve.

- **Opponens digiti minimi**

Origin: From the flexor retinaculum and adjacent carpal bones.

Insertion: To the shaft of the fifth metacarpal.

Action: Opposition of the little finger.

Nerve supply: Ulnar nerve.

- **The lumbrical muscles**

Origins: The four muscles arise from the sides of the tendons of the flexor digitorum profundus.

Insertions: To the lateral sides of the dorsal extensor expansions of the extensor digitorum tendon.

Actions: Flexion of the metacarpophalangeal joints and extension of the interphalangeal joints.

Nerve supply: Medial two muscles by the ulnar nerve and lateral two by the median nerve.

- **The interossei**

Origins: *Dorsal* from adjacent sides of four metacarpals and *palmar* from one side of each metacarpal.

Insertions: Both dorsal and palmar are inserted into the sides of the proximal phalanges and the dorsal extensor expansions in such a way that the dorsal interossei abduct the fingers and the palmar adduct them (Fig. 44.2).

Actions: Adduction and abduction as above. Both sets of muscles produce flexion of the metacarpophalangeal joints and extension of the interphalangeal joints, as in the 'precision grip'.

Nerve supply: All by the deep branch of the ulnar nerve.

The lower limb

- **Gluteus maximus**

Origin: From the posterior part of the gluteal surface of the ilium, the back of the sacrum and its associated ligaments.

Insertion: To the gluteal tuberosity of the femur (25%) and the iliotibial tract (75%).

Actions: Extension and lateral rotation of the thigh.

Nerve supply: Inferior gluteal nerve.

- **Gluteus medius**

Origin: From the gluteal surface of the ilium.

Insertion: To the greater trochanter.

Actions: Abduction and medial rotation of the thigh.

Nerve supply: Superior gluteal nerve.

- **Gluteus minimus**

Origin: From the gluteal surface of the ilium below gluteus medius.

Insertion: To the greater trochanter.

Actions: Abduction and medial rotation of the thigh. The most important action of medius and minimus is to prevent the pelvis tilting to the unsupported side when taking the weight on one leg, as in walking.

Nerve supply: Superior gluteal nerve.

- **Tensor fasciae latae**

Origin: From the anterior part of the crest of the ilium.

Insertion: To the lateral condyle of the tibia via the iliotibial tract.

Actions: Extension of the knee joint. Helps the gluteal muscles to prevent tilting of the pelvis.

Nerve supply: Superior gluteal nerve.

- **Piriformis**

Origin: From the front of the sacrum.

Insertion: Into the greater trochanter via the greater sciatic notch.

Action: Lateral rotation of the thigh.

Nerve supply: From the sacral plexus.

- **Obturator internus**

Origin: From the inner surface of the hip bone and the obturator membrane.

Insertion: To the greater trochanter via the lesser sciatic notch.

Action: Lateral rotation of the thigh.

Nerve supply: From the sacral plexus.

- **Quadratus femoris**

Origin: From the outer surface of the ischial tuberosity.

Insertion: To the quadrate tubercle on the intertrochanteric crest of the femur.

Action: Lateral rotation of the thigh.

Nerve supply: From the sacral plexus.

- **Obturator externus**

Origin: From the obturator membrane and the surrounding bone.

Insertion: To the trochanteric fossa of the femur.

Action: Lateral rotation of the thigh.

Nerve supply: Obturator nerve.

- **Iliacus**

Origin: From the concave inner surface of the ilium.

Insertion: Passes under the inguinal ligament to the lesser trochanter, in company with psoas major.

Action: Flexion of the thigh.

Nerve supply: Femoral nerve.

- **Psoas major**

Origin: From the transverse processes and the sides of the bodies and intervertebral discs of the lumbar vertebrae.

Insertion: Passes under the inguinal ligament to the lesser trochanter in company with iliacus. (Joint muscle often called *iliopsoas*.)

Action: Flexion of the thigh.

Nerve supply: From the lumbar plexus.

- **Sartorius**

Origin: From the anterior superior iliac spine.

Insertion: To the medial side of the upper end of the tibia just in front of gracilis and semitendinosus.

Actions: Flexion and abduction of the thigh and flexion of the knee (the 'tailor's position').

Nerve supply: Femoral nerve.

- **Quadriceps femoris:**
  - **Rectus femoris**

  Origin: From the anterior inferior iliac spine (*straight head*) and the upper lip of the acetabulum (*reflected head*).
  - **Vastus medialis**

  Origin: From the medial lip of the linea aspera.
  - **Vastus lateralis**

  Origin: From the lateral lip of the linea aspera.
  - **Vastus intermedius**

  Origin: From the lateral and anterior surfaces of the femur.

Insertion of quadriceps: The four parts of quadriceps are inserted into the patella and, from here, to the tubercle of the tibia. The patella is thus a *sesamoid bone* in the tendon of quadriceps.

Actions: Extension and stabilization of the knee. Rectus is also a weak flexor of the thigh.

Nerve supply: Femoral nerve.

- **Pectineus**

Origin: From the superior ramus of the pubis.

Insertion: To the back of the femur between the lesser trochanter and the linea aspera.

Actions: Adduction and flexion of the thigh.

Nerve supply: Femoral and obturator nerves.

- **Adductor longus**

Origin: From the front of the pubis just below the pubic tubercle.

Insertion: To the middle third of the linea aspera.

Action: Adduction of the thigh.

Nerve supply: Obturator nerve.

- **Adductor brevis**

Origin: From the inferior ramus of the pubis.

Insertion: To the upper part of the linea aspera.

Action: Adduction of the thigh.

Nerve supply: Obturator nerve.

- **Adductor magnus**

Origin: From the inferior ramus of the pubis and the ramus of the ischium, back as far as the ischial tuberosity.

Insertion: To the whole length of the linea aspera and to the adductor tubercle of the femur.

Actions: Adduction and extension of the thigh (the latter action is carried out by the 'hamstring' part of the muscle which arises from the ischial tuberosity).

Nerve supply: Adductor part by the obturator nerve and hamstring part by the sciatic nerve.

- **Gracilis**

Origin: From the inferior ramus of the pubis and the ramus of the ischium.

Insertion: To the medial side of the tibia between sartorius and semi-tendinosus.

Action: Adduction of the thigh.

Nerve supply: Obturator nerve.

- **Biceps femoris**

Origin: Long head from the ischial tuberosity and short head from the linea aspera.

Insertion: By a thick tendon into the head of the fibula.

Actions: Extension of the hip and flexion of the knee.

Nerve supply: Sciatic nerve (both components).

- **Semitendinosus**

Origin: From the ischial tuberosity.

Insertion: To the medial side of the front of the tibia, behind sartorius and gracilis.

Actions: Extension of the hip and flexion of the knee joint.

Nerve supply: Sciatic nerve (tibial component).

- **Semimembranosus**

Origin: From the ischial tuberosity.

Insertion: To a groove on the tibial medial condyle.

Actions: Extension of the hip and flexion of the knee.

Nerve supply: Sciatic nerve (tibial component).

- **Tibialis anterior**

Origin: From the lateral surface of the tibia.

Insertion: To the base of the first metatarsal and the medial cuneiform.

Actions: Dorsiflexion and inversion of the foot.

Nerve supply: Deep peroneal nerve.

- **Extensor hallucis longus**

Origin: From the middle third of the shaft of the fibula.

Insertion: To the base of the distal phalanx of the big toe.

Actions: Extension of the big toe and dorsiflexion of the foot.

Nerve supply: Deep peroneal nerve.

- **Extensor digitorum longus**

Origin: From the shaft of the fibula.

Insertion: To the bases of the middle and distal phalanges of the four lateral toes via the dorsal extensor expansions.

Actions: Extension of the toes and dorsiflexion of the foot.

Nerve supply: Deep peroneal nerve.

- **Peroneus tertius**

Origin: Formed by the lower part of extensor digitorum longus.

Insertion: Into the base of the fifth metatarsal.

Action: Dorsiflexion of the foot.

Nerve supply: Deep peroneal nerve.

- **Extensor digitorum brevis**

Origin: From the upper surface of the calcaneus.

Insertion: To the proximal phalanx of the big toe and to the extensor digitorum longus tendons of the next three toes.

Actions: Dorsiflexion of the foot and extension of the toes.

Nerve supply: Deep peroneal nerve.

- **Peroneus longus**

Origin: From the upper two-thirds of the lateral surface of the shaft of the fibula.

Insertion: To the base of the first metatarsal and the medial cuneiform, via the groove on the cuboid.

Action: Eversion of the foot.

Nerve supply: Superficial peroneal nerve.

- **Peroneus brevis**

Origin: From the lower two-thirds of the shaft of the fibula.

Insertion: To the base of the fifth metatarsal.

Action: Eversion of the foot.

Nerve supply: Superficial peroneal nerve.

- **Gastrocnemius**

Origin: From the femur just above both femoral condyles.

Insertion: To the middle third of the back of the calcaneus via the tendo calcaneus.

Actions: Plantar flexion of the foot; weak flexion of the knee.

Nerve supply: Tibial nerve.

- **Soleus**

Origin: From the soleal line of the tibia and the upper part of the back of the fibula.

Insertion: To the middle third of the back of the calcaneus via the tendo calcaneus in common with the gastrocnemius.

Actions: Plantar flexion of the foot. It is the main factor in the 'muscle pump'.

Nerve supply: Tibial nerve.

- **Plantaris**

A detached piece of the lateral head of gastrocnemius with similar properties.

- **Popliteus**

Origin: From the back of the tibia above the soleal line.

Insertion: To the lateral condyle of the femur.

Actions: Flexion and medial rotation of the leg (thus 'unlocking' the extended knee joint).

Nerve supply: Tibial nerve.

- **Tibialis posterior**

Origin: From the back of the tibia and fibula.

Insertion: To the tuberosity of the navicular and other tarsal bones.

Actions: Plantar flexion and inversion of the foot.

Nerve supply: Tibial nerve.

- **Flexor hallucis longus**

Origin: From the back of the fibula.

Insertion: To the base of the distal phalanx of the big toe.

Action: Flexion of the big toe.

Nerve supply: Tibial nerve.

- **Flexor digitorum longus**

Origin: From the back of the tibia.

Insertion: To the bases of the distal phalanges of the four lateral toes via the openings in the tendons of flexor digitorum brevis.

Action: Flexion of the toes.

Nerve supply: Tibial nerve.

- **Abductor hallucis**

Origin: From the calcaneal tuberosity.

Insertion: To the medial side of the proximal phalanx of the big toe.

Action: Abduction of the big toe.

Nerve supply: Medial plantar nerve.

- **Flexor digitorum brevis**

Origin: From the calcaneal tuberosity.

Insertion: To the sides of the middle phalanges of the lateral four toes, its tendons being perforated by those of flexor digitorum longus.

Action: Flexion of the toes.

Nerve supply: Medial plantar nerve.

- **Abductor digiti minimi**

Origin: From the calcaneal tuberosity.

Insertion: To the proximal phalanx of the little toe.

Actions: Flexion and abduction of the little toe.

Nerve supply: Lateral plantar nerve.

- **Lumbricals**

Origin: From the tendons of flexor digitorum longus.

Insertion: To the dorsal extensor expansions.

Actions: Assist the actions of the interossei.

Nerve supply: First lumbrical by the medial and the others by the lateral plantar nerves.

- **Flexor digitorum accessorius**

Origin: From the undersurface of the calcaneus.

Insertion: Into the side of the tendon of flexor digitorum longus.

Action: Tenses the tendon of this muscle.
Nerve supply: Lateral plantar nerve.

- **Flexor hallucis brevis**

Origin: From the underside of the cuboid.
Insertion: To the sides of the proximal phalanx of the big toe with a sesamoid bone in each tendon.
Action: Flexion of the proximal phalanx of the big toe.
Nerve supply: Medial plantar nerve.

- **Adductor hallucis**

Origin: From the heads and from the bases of the metatarsals.
Insertion: Into the lateral side of the proximal phalanx of the big toe.
Action: Adduction of the big toe.
Nerve supply: Lateral plantar nerve.

- **Flexor digiti minimi brevis**

Origin: From the base of the fifth metatarsal.
Insertion: Into the proximal phalanx of the little toe.
Action: Flexion of the little toe.
Nerve supply: Lateral plantar nerve.

- **Interossei:**
  - **Dorsal**

Origin: From adjacent sides of the metatarsals.
Insertion: Into the dorsal extensor expansions and the sides of the proximal phalanges.
Actions: Abduction of the toes; flexion of the metatarsophalangeal joints and extension of the interphalangeal joints.
Nerve supply: Lateral plantar nerve.

  - **Plantar**

Origin: From the bases of three of the metatarsals.
Insertion: Into the dorsal extensor expansions and the sides of the proximal phalanges.
Actions: Adduction of the toes and assisting the other actions of the dorsal interossei.
Nerve supply: Lateral plantar nerve.

The head and neck
The muscles of mastication
- **Temporalis**

Origin: From the lateral side of the skull below the temporal line.
Insertion: To the coronoid process of the mandible, extending down the anterior border as far as the third molar tooth. It passes deep to the zygomatic arch.
Actions: Closes the mouth and clenches the teeth. The posterior fibres are horizontal and help to retract the mandible when closing the mouth.

- **Masseter**

Origin: From the lower border of the zygomatic arch.
Insertion: To the lateral side of the mandible in the region of the angle.
Actions: Closes the mouth and clenches the teeth.

- **Lateral pterygoid**

Origin: From the lateral pterygoid plate.
Insertion: To the neck of the mandible and the intra-articular disc of the temporomandibular joint.
Actions: It protrudes the mandible and moves the head of the mandible onto the articular eminence when the mouth is opened.

- **Medial pterygoid**

Origin: From the lateral pterygoid plate.
Insertion: To the medial surface of the mandible near the angle.
Action: Helps to close the mouth.

Nerve supply of the muscles of mastication: The muscles of mastication are all supplied by the mandibular division of the trigeminal nerve.

The muscles of facial expression
The principal muscles of facial expression are shown in Fig. 72.1, from which their actions can be deduced. Only a few of them will be described here.

- **Occipitofrontalis**

Origin: The *frontal belly* extends backwards from the forehead region and the *occipital belly* extends forwards from the occipital bone.
Insertion: Both parts are inserted into the galea aponeurotica (epicranial aponeurosis).
Actions: The frontal belly lifts the eyebrows when looking upwards and both bellies can move the whole scalp on the underlying loose fascia.

- **Orbicularis oculi**

In two parts. The orbital part surrounds the whole eye, blending with the frontal belly of occipitofrontalis. It closes the eye tightly. The palpebral part is in the eyelid and closes the eye gently as in sleep.

- **Orbicularis oris**

Surrounds the whole mouth and blends in with the surrounding muscles that are shown in Fig. 72.1. Closes the mouth and protrudes the lips.

- **Buccinator**

Origin: From the pterygomandibular ligament, where it is continuous with the superior constrictor of the pharynx.
Insertion: Blends in with the orbicularis oris. Tightens the cheeks and keeps them in contact with the gums, thus preventing food collecting in the vestibule of the mouth.

- **Platysma**

Origin: From the skin over the upper part of the chest, crossing the mandible to blend in with the orbicularis oris. Pulls down the corners of the mouth and has an anti-sphincteric action on the neck as in loosening a tight collar.

Nerve supply of the muscles of facial expression: All the muscles of facial expression are supplied by the facial nerve.

- **Trapezius**

Origin: From the superior nuchal line, the ligamentum nuchae and the spines of all the thoracic vertebrae.
Insertion: Into the spine of the scapula, the lateral edge of the acromion and the lateral third of the clavicle.
Actions: Extends the head on looking upwards, raises the tips of the shoulders, braces the shoulders back, and helps serratus anterior to rotate the scapula during abduction of the arm.
Nerve supply: Spinal accessory nerve.

- **Sternocleidomastoid**

Origin: From the front of the manubrium (by a narrow rounded tendon) and the medial third of the clavicle.
Insertion: To the mastoid process and the lateral part of the superior nuchal line.
Actions: Rotates the head to the opposite side and flexes the cervical spine to the same side, thus bringing the ear nearer to the shoulder of the same side. Both muscles acting together can flex the cervical spine against resistance. An accessory muscle of respiration.
Nerve supply: Spinal accessory nerve.

- **Scalenus anterior**

Origin: From the anterior tubercles of the transverse processes of several cervical vertebrae.
Insertion: To the scalene tubercle on the medial border of the first rib, by means of a narrow pointed tendon.

Actions: Flexes the cervical spine, produces lateral flexion to the same side and rotation to the opposite side. An accessory muscle of respiration.

Nerve supply: Cervical spinal nerves.

• **Scalenus medius**

Origin: From the posterior tubercles of the transverse processes of most of the cervical vertebrae.

Insertion: Into a large area on the first rib behind scalenus anterior.

Actions: Similar to those of scalenus anterior.

Nerve supply: Cervical spinal nerves.

• **Other muscles**

Other muscles of the neck, such as the strap muscles, are described fully in the text.

Index

Note: Italicized page numbers refer to figures